Understanding the Recent Discovery of Noah's Ark

Understanding the Recent Discovery of Noah's Ark

and how the evidence of Noah's Flood was reimagined as a long Ice Age!

Philip Ernest Williams

CLSP

Rev 1.1 June 2025

ISBN 978-0-9793102-6-3

Library of Congress Control Number: 2024920695

Published by

Christian Leaders & Scholars Press

Cover Photo: author inspecting Noah's Ark buried on Mount Ararat

Preface

Having spent his childhood in Ottoman era Turkey, the founding editor of the *National Geographic* magazine Gilbert H. Grosvenor would have learned that the natives of eastern Turkey believe that Noah's Ark still rests on Mount Ararat. This might explain Grosvenor's interest in the world's fascinating local cultures and why *National Geographic* regularly covered reports of Noah's Ark including the announcement in 2010 about the site that I visited in 2014. The latter is reported to have been the most popular digital story ever covered by *National Geographic*.

When interviewed by the famous Ark Searcher Eryl Cummings, Grosvenor explained the significance of such an unlikely discovery,

"If the Ark of Noah is ever found,

it will be the greatest archaeological discovery in history,

the greatest event since the resurrection of Christ,

and will alter all the currents of scientific thought."

No one expected this and no one was prepared for what they could hardly believe, what we have yet to understand, which is the purpose of this book.

In Part I, I review this recent discovery and cover it in detail in Part V after I explain the archaeological context which begins addressing the third of Grosvenor's predictions: how the discovery will alter all the currents of scientific thought. Another way it will do the latter is by bringing attention to how our world has overlooked the massive evidence of Noah's Flood due to geologists having invented a long Ice Age to explain the deposits of a one-year Flood, thus artificially creating what is now understood as the great antiquity of man.

After reading this book, I trust that you will be impressed with just how much the light of Noah's Ark and Flood simplifies our understanding of the ancient past which currently exists in such a fragmented, muddled, and ever-changing state that no one person can completely know what is being taught much less understand how and why these beliefs came to be accepted as parts of the scientific narrative. Much of this book is a summary of the current scientific narrative of origins, which has perhaps never before been summarized in such detail. Many will find this book a valuable resource if only for understanding the current scientific narrative. However, a far simpler understanding of the origins of mankind begins with the discovery of the archaeological evidence of Noah's Flood as summarized in Part II, which led to my coming to the defense of the 2008 discovery on Mount Ararat.

The chief sciences forming the scientific narrative of origins developed in the following historical order:

1. *Chronology:* The calendrical dating of the ancient world
2. *Cosmogony:* The development of the earth and physical heavens
3. *Anthropology:* The study of mankind and his origins
4. *Philology:* The study of ancient texts, languages, and scripts
5. *Geology:* The study of the earth and its origins
6. *Archaeology:* The study of the artifacts and remains of humans
7. *Evolutionary Biology:* The history of living things

Though the historical development of these sciences can hardly be covered in a single book much less one as short as this, Part II covers the profound mistakes that have been made in the foundations of archaeology which have hidden the evidence of those who drowned in the Flood. Part III explains how geological science has hidden the evidence of the Flood by inventing an Ice Age. Part IV explains how the idea of an Ice Age was responsible for artificially creating the vast antiquity of man that so challenged the Bible. Part V discusses the recent discovery of what appears Noah's Ark. Part VI explains how the waters of the Flood are the same as those used for the Creation of the heavens and the earth. Part VII summarizes the new school or origins outlined in this book which I refer to as *Scripture Science* to distinguish it from what is called *Creation Science*. Part VIII shows how returning to the biblical narrative of origins in the light of this recent discovery will change the world in the profound way that Grosvenor predicted.

While the reader will easily understand the thesis of this book given in its title, absorbing the vast number of topics in the subsequent sections will challenge most readers. I must however cover these topics not only to demonstrate the profound historicity of Noah's Ark and Flood but also to explain how the evidence of this event has been hidden from modern eyes by certain falsehoods underlying the foundations of the historical sciences. However, after reading the Introduction in Part I you may prefer to review the Table of Contents and skip to chapters that particularly interest you, perhaps those of Part V which cover the remains of the Ark. After this, you can return to the more difficult chapters which address the complex and fragmented teachings of the current historical sciences. I trust that it will change the way that you understand science, history, and our current world.

My confidence in the trustworthiness of the Scriptures constrains how I see the world and understand its history. Thus, the reader will recognize that I write boldly and unapologetically not just as a Christian but as a born-again Christian who reads the Scriptures in the light of Christ. Not only do I read the Scriptures in a completely literal way but also in the premodern way that the Scriptures are written, which is different from how modern Christians read them. I write first to those who share the presuppositions of our common faith, then to all who may be interested in this discovery and a simpler and more integrated understanding of origins.

I wish however to express my indebtedness to the many non-Christian scholars and scientists who seek and publish truth, from whom I have learned much. Christians should know that those of us in the West borrowed our earliest science from the Islamic Arabs who bequeathed us their algebra and Arabic numerals. Devout Muslims and Jews also share belief in the historicity of Noah's Ark and Flood. Indeed, everyone should be interested in their origins. May all who read this book work together to make this world a better place. Discovering the truth about our common origins will prepare the way for just such a future.

Philip Ernest Williams　　　　　　　　　　Charlotte, NC

TABLE OF CONTENTS

List of Photos, Illustrations, and Credits

Part I

Introduction

1

An Unexpected Announcement

In April 2010, I was about to publish the first ever archaeological explanation of Noah's Flood[1] when explorers from Hong Kong and Turkey announced the discovery of what they believe are the remains of Noah's Ark. These remains lie buried under volcanic rock and ice near the top of Mount Ararat in far eastern Turkey. The Turkish name for this mountain is *Agri Dagh*. It rises high above the surrounding mountains in the land of the ancient Kingdom of Urartu, which English Bibles translate as *Ararat*.

Fig 1 – My friend Panda Lee points to the buried wooden ship

Those who have lived on this mountain from times immemorial have always believed these to be the remains of Noah's Ark. But concerned that American Ark searchers with GPS technology might discover and exploit their secret, the natives decided that it was time to share the location with explorers whom they trusted.

[1] Williams, Philip Ernest, *The Archaeological Evidence of Noah's Flood* (2011, CLS Press), summarized in Part II. Hereafter, I may refer to this book as the *Archaeological Evidence*.

Whatever be the truth, this is the one tiny spot on our large earth's surface that matches the landing site of the Ark as given in the Bible. The passage in the Bible does not mention the specific mountain where the Ark rested but context is important. The surrounding mountain tops would have been covered by water only if the Ark had rested near the top of this highest mountain in all Ararat!

And the ark rested in the seventh month, on the seventeenth day of the month, upon the mountains of Ararat. And the waters decreased continually until the tenth month: in the tenth month, on the first day of the month, were the tops of the mountains seen. [Genesis 8:4-5]

Fig 2 - Viewing the mountains of Ararat from near the site of the Ark

While the explorers sought an archaeological investigation of their discovery, I knew this would be difficult because current science has failed to understand the evidence or even the possibility of a worldwide Flood. A Flood that could float this ship to the top of a mountain that is more than three miles above sea level must have necessarily covered the earth.

Young Earth Creationists do teach a worldwide Flood. However, this school of origins misunderstand the geological evidence of Noah's Flood which our current science has mistaken as the deposits of a supposed long Ice Age. As a result, Creationists lack an archaeological understanding of both the Flood and the Ark. They can neither identify the remains and civilizations of those who drowned in the Flood nor trace the migrations of the historical nations from the Ark's landing site. These Creationists do not seem troubled by the fact that more than ninety five percent of the victims of the Flood which they do identify are fish, the very opposite of the account in the Bible!

***Every living thing that moved on land perished**—birds, livestock, wild animals, all the creatures **that swarm over the earth,** and all mankind**. Everything on dry land that had the breath of life in its nostrils** died. Every living thing on the face of the earth was wiped out; people and animals and **the creatures that move along the ground and the birds** were wiped from the earth. Only Noah was left, and those with him in the ark. [Gensis 7:21-23]*

Fish can swim and breath in water. Clearly, the "Flood" of the Creationist is not the Flood of Noah. Because Mount Ararat sits on what these Creationists claim as the evidence of the Flood, they do not believe this mountain even existed at the time of the Flood. Though joined in opposition by Old Earth Creationists who teach a local Flood, Creationists ironically became the chief opponents of the discovery of Noah's Ark!

Thus, I became the discovery's chief defender against charges of hoax which immediately followed the announcement. In June 2010, I arranged for the Hong Kong explorers to present their findings to a group of scientists, archaeologists, professors, and university officials gathered at a private meeting at a Charlotte, NC science museum. Later that year, the explorers presented their evidence to the 2010 *National Conference on Christian Apologetics*. In January 2012, I traveled to Turkey to meet the discoverer, Ahmet Ertugrul, better known as Parasut (Turkish for *parachute*). Accompanied by Sheila Bishop, a trustee of the venerable *American Schools of Oriental Research,* I visited this ancient wooden structure on November 9, 2014 for the purpose of obtaining American archaeological involvement. I took the photo of Sheila and Parasut (Fig 2) on the day that we visited the second deck of Noah's Ark.

In 2015, my work on origins received the endorsement of the late Dr. Norman Geisler, the great pioneer of evangelical apologetics. In October 2015, Dr. Geisler and I appeared on FOX News as Dr. Geisler explained how the evidence presented by the explorers matches the biblical account of Noah's Ark.

Having studied paleontology and the history of geology, Dr. Geisler was aware that what are currently taught as the deposits of supposed Ice Age glaciers were initially understood as evidence of what was perhaps a worldwide Flood. This was the standard geological view of the founders of geological science for explaining deposits which they accordingly called *Diluvium,* though later renamed *Pleistocene* after having been re-envisioned as deposits of Ice Age glaciers. As I was completing my research, I discovered that these same deposits cover the prehistoric burial mounds that archaeologists have misunderstood as elaborate religious burials. As I will explain in Chapter 8, these are manifestly victims of a mass drowning

Dr. Geisler encouraged me to complete the research on just how and why the deposits of Noah's Flood were re-envisioned as a long Ice Age, thereby pushing back the beginning of mankind from about six thousand to hundreds of thousands of years and just in time for Darwin's theory of evolution. The resulting great antiquity of man has been an even greater challenge to the truth of the Bible than evolution. Evolution or not, such great antiquity of man challenges the history of man given in the Bible which begins with Adam. This is because the redemption from sin by the blood of Jesus is a consequence of the sin of Adam. Hence, this challenge to the historicity of Adam also challenges the very truth of the gospel.

In Part III, I will explain the scarcely understood history of how the deposits of Noah's Flood were re-envisioned as having been spread by glaciers of a newly invented Ice Age. Understanding how these supposed Ice Age deposits were from Noah's Flood is also essential for understanding the recent origins of man as well as the remains of Noah's Ark that I visited near the top of Mount Ararat. The revelation of how the evidence of Noah's Flood has been hidden from modern eyes by a supposed Ice Age will vanquish the scientific challenge to the Bible that has for the last two centuries anchored unbelief in what were formerly Christian nations.

2

Mankind's Origins in the Light of Modern Science

According to the scientific narrative that developed from confusing the deposits of the Flood with a long Ice Age, we humans began our life on earth some place in Africa during the Ice Age about 300,000 years ago. Whether this happened through evolution or special creation, we have spent the vast majority of our history roaming across the earth in small bands to hunt wild game and gather whatever food grows in the wild. The Ice Age ended about 12,000 years ago but it was not until several thousand years after the Ice Age that we learned to domesticate the wild animals that we followed and cultivate the wild grasses that we gathered. This agriculture revolution was followed a few thousand years later by an urban revolution when we learned to make tools and weapons of bronze and learned the way of trade so as to live in cities.

Fig 3 - Cavemen of the Ice Age

So entrenched has become this scientific narrative that it seems never to occur to those who have been taught this narrative to seriously question it. Our modern world has come to regard it as firmly established truth. This contrasts with the great attention given to the matter of evolution. Christians may also fail to notice how this narrative of early man also conflicts with the history of agriculture and civilization given in the Bible which began almost simultaneously with God's creation of man. These developments leading to human civilization are believed to have taken place in the *Ancient Near East,* which were the lands of the Bible!

According to the scientific narrative, civilization was preceded by an agriculture revolution in what is called the *Fertile Crescent* which extends from southern Iraq, through Syria and Israel down into Egypt. Writing was invented in the world's first cities which also arose in the Fertile Crescent. As now believed, this took place in what is now Iraq around the time of Noah's Flood as can be calculated from the Bible. However the scientific account of man's history recognizes no evidence of this worldwide Flood though it has been vaguely remembered from ancient times by nations across the earth.

In the late nineteenth century archaeologists exploring the lands of the Ancient Near East found texts on clay tablets which tell of a Flood with many similarities to the one in the Bible. Some scholars believe that stories about a great Flood arose due to the melting of glaciers at the end of the Ice Age. Such floods are supposed as having been local floods which may have ended the limited world of those who lived in the river valleys where these floods occurred.

According to critical scholars, the account of the Flood in the Bible was borrowed from the "more advanced" nations of the Ancient Near East. These scholars tell us that the final editing of Genesis where is written the record of mankind's beginning took place during the Babylonian captivity of the Jews sometime after the eighth century before Christ. The sources and editors of the book of Genesis are referred to by the letters JEPD which stand for J (the Yahwist source), E (the Elohist source), P (the Priestly editor), and D (the Deuteronomist editor), the latter editor supposed to have edited the writings of the Bible during the time of King Josiah so as to require a single place of worship.

While archaeologists have since discovered that the archaeology of the lands of the Bible accords with the history written in the Bible from the time of Abraham, the same has not been the case for the earliest chapters of Genesis. Rather, the modern narrative of the origins of mankind outlined above challenges the essential message of the Bible including the very words of Jesus who referred to these earliest chapters of Genesis more than anyone else in the Bible.

I began addressing the scientific challenge to the truth of the Bible with the publication of my 2011 book, the *Archaeological Evidence of Noah's Flood*. However, the recent discovery on Mount Ararat cannot be understood until we understand the true geological evidence of Noah's Flood which the current science believes to have been deposited by the glaciers of a long Ice Age.

3

The Mistaken and True Geological Evidence of the Worldwide Flood

Young Earth Creationists challenge this scientific narrative of origins by rearranging the chronology of the mainstream narrative to accord with an earth that was created in less than one week and which is only six to ten thousand years old. Regardless their good and pious aims, because Creationists just rearrange the dating of events in the narrative of mainstream science, they are beholden to the scientific narrative of the Ice Age. Creationists however use an earth-destroying Flood to explain the presence of fossils in the earth's deepest rocks which they claim were buried by Noah's Flood.

A problem is that Creationists can find no evidence of the remains of humans or their houses and artifacts in these deeply buried rocks where exists petrified fossils, many being species such as dinosaurs that have long been extinct. There should however be plentiful evidence of artifacts of stone, brick, and bronze from the antediluvian civilization. If humans could somehow have outrun the rising waters of the Flood so as not to have become deeply buried, their stone, brick, or wooden houses could not have done so. Moreover, stone and bronze artifacts are harder and more enduring than those of fish and soft dinosaur bones!

Ironically, Creationists agree with mainstream science's argument for an Ice Age, though they see the Ice Age as having taken place after the Flood when, according to them, dinosaurs were still roaming the earth. But Creationists cannot find the remains of dinosaurs in the deposits attributed to the Ice Age nor, more important, find the remains of the people and antediluvian settlements and civilization drowned by the Flood. This is because the deeper layers of the earth long predate the era of man who, as I will explain, probably does date from only about six thousand years ago just as taught in the Bible.

Dating the first humans so much earlier was due to geologists having imagined a very long Ice Age to explain away the true deposits of Noah's Flood which the founders of geological science had identified in the early nineteenth century! These founders of geological science, the same as determined the earth to be far older than man, not only discovered the evidence of the sudden creation of animals that is in line with the order of Creation in the first chapter of Genesis, but they also discovered the geological deposits of a worldwide Flood which they accordingly referred to as the *Diluvium*.

As should be expected from a Flood that took place during the era of man, Diluvium deposits are found near the surface of the earth. The Diluvium also contains the fossils of extinct animals, though ones similar to living species. These recent fossils are buried in deposits of loose gravel, sand, clay, rocks, and silt that lie just below the topsoil of unimproved lands across the earth that have not been eroded by ordinary floods and present-day rivers and streams. The Diluvium deposits are themselves from a great erosion that took place just less than 4,500 years ago as the result of the worldwide Flood of Noah. As one might expect, this worldwide Flood eroded soil, gravel, and rocks from the surface of the mountains and continents depositing them in valleys and river deltas while covering the houses, cities, and fortifications of what are now understood as prehistoric civilizations.

That the deposits of Noah's Flood cover what are now understood as prehistoric civilizations means that we do not need to guess concerning the effects of the Flood. Rather, we can study them as in the case of those Stone Age dwellings buried by the deposits of the Flood shown in Fig 4. Not surprisingly, we discover that the prehistoric archaeological remains in mountains and highlands are scarcely buried. But, as we might expect in the case of low-lying areas, these houses, fortifications, and cities appear as burial mounds, kurgans, or archaeological tels, having been assigned these different names by archaeologists working in different

Fig 4 – A Stone Age dwelling buried by deposits from Noah's Flood

regions of the world. Most of these remains are too deeply buried for inexpensive excavation. Others lie offshore at the bottom of the oceans. This happened due to the rising of the oceans as water once held in great reservoirs just below the surface of the earth flowed into the oceans as the Flood waters rose upon the earth. While the continents and mountains rose due to the emptying of the weight of these underground waters, the oceans stretched and deepened as recalled in this famous Psalm.

He set the earth on its foundations; it can never be moved. You covered it with the watery depths as with a garment; the waters stood above the mountains. But at your rebuke the waters fled, at the sound of your thunder they took to flight; they flowed over the mountains, they went down into the valleys, to the place you assigned for them. You set a boundary they cannot cross; never again will they cover the earth. [Psalm 104: 5-9]

Again, no one should confuse the true evidence of the biblical Flood with the supposed geological evidence of the Flood now being taught by Young Earth Creationists. According to these Creationists, the Flood destroyed not just the earth's surface but miles deep of bedrock so as to bury now extinct forms of life. That is contrary to what the Bible teaches! That the book of Genesis mentions historical lands of Cush and Havilah and the Tigris and Euphrates Rivers as existing before the Flood is a problem for Creationists! This is because these rivers flow above the deeper rocks and deposits that Young Earth Creationists attribute to the Flood. Young Earth Creationists thus suppose that these landmarks from before the Flood had to be re-created after the Flood!

An Ark of steel could not have survived such great destruction of the earth as taught by these Creationists. In truth, with the exception of rocks broken and shot forth to release the springs of the deep, the waters of the Flood eroded the surface of the earth only above solid bedrock. Had the Ark been built on a modest elevation, the rising waters would have gently lifted the Ark.

Because this current generation knows only the account of the worldwide Flood taught by the Young Earth Creationists, it is essential that I address their strange understanding of the Flood in this introduction to the true narrative of the worldwide Flood of Noah as taught in the Bible. Otherwise, most will suppose that I am confirming these teachings that have so confused the current world about what the Bible teaches.

Know however that the chief purpose of this book is not to challenge Young Earth Creationist teachings, but rather the teachings of mainstream science that I discussed in the previous chapter. It gives me no pleasure to challenge the one school of origins which until the explanation put forth in this book defended a worldwide Flood and thus is now embraced by most Bible-believing Christians. So popular have become their teachings in the last few decades that they are teaching unbelievers how to read the Bible, thus affirming the rejection of the Bible by those trained in the sciences. Not only do Creationists fail to understand the evidence of the worldwide Flood but they also help blind eyes to the true evidence of the Flood.

However, the same is true of Old Earth Creationists who teach a local Flood. I trust that both Young and Old Earth Creationists will consider what I put forth here as the true geological evidence of the worldwide Flood and my answers to the theological arguments for a young earth which, as I explain below, are more in line with the theological and exegetical arguments of Old Earth Creationists. I promised to treat the responses of both schools to what I put forth here with respect.

Young Earth Creationists believe they can decide the age of the earth strictly from reading the Bible apart from any observation and study of God's Creation. This differs from my method, which is to study the Creation in the light of the Bible. God has given us this light. We ought to use it. This is what I call *Scripture Science* which differs from the presuppositional hermeneutics of the Young Earth Creationists. In other words, I do not believe the Bible can be understood apart from believing and applying what it teaches.

BAD THEOLOGICAL AND HERMENEUTICAL ARGUMENTS FOR A YOUNG EARTH

Whatever scientific challenges Young Earth Creationists face, they have developed theological and hermeneutical arguments to promote their teachings among Christians like me who are committed to a literal reading of the Scriptures. But these are not sound theological arguments! For example, Creationists interpret the Apostle Paul as having taught that no animals died before Adam sinned! Thus, according to them, no fossils of dead animals should predate the evidence of man. That however is to misread the Apostle Paul. Because Adam sinned, death came to humans!

Therefore, just as sin entered the world through one man, and death through sin, and in this way death came to all people, because all sinned. [Romans 5:12]

Before Adam sinned, God had commanded the animals to fill the earth. However, there was never any danger that potentially immortal rabbits would overrun the Garden of Eden before Adam sinned. This is because the fruit of the Tree of Life was not for animals. Death will be defeated in the New Creation that began with the resurrection of Jesus. In the New Creation, the nature of animals will be changed such that the lion will lay down with the lamb. The New Creation however is not a restoration of the original Creation as some theologians including Young Earth Creationists wrongly teach.

God proclaimed his first Creation good. As the Apostle Paul tells us in the first chapter of Romans, we observe God's providentially good and beautiful Creation in the great outdoors even during this time of sinful humanity. Death is the result of God's wise rule of his first creation. Whether by God or man, animal populations must be managed as God has also instructed us:

Then God said, "Let us make mankind in our image, in our likeness, so that they may rule over the fish in the sea and the birds in the sky, over the livestock and all the wild animals and over all the creatures that move along the ground" [Gen 1:26]

God did not create the animals in his image. Some animals were created to be eaten as the fulfilment of their purpose. Such is a good and noble death for animals, some of which serve as a type of Christ's sacrifice when he died for our sins. The death of animals is part of the order of God's first Creation which will one day pass away (not restored!). This is why we so look forward to his New Creation. Though God-denying evolutionists place humans on the same level as animals, Christians should not.

Another Creationist argument for a young earth is based on one of the ten commandments:

Remember the Sabbath day by keeping it holy. Six days you shall labor and do all your work, but the seventh day is a sabbath to the Lord your God. On it you shall not do any work, neither you, nor your son or daughter, nor your male or female servant, nor your animals, nor any foreigner residing in your towns. For in six days the Lord made the heavens and the earth, the sea, and all that is in them, but he rested on the seventh day. Therefore, the Lord blessed the Sabbath day and made it holy. [Exodus 20:8-11]

Here Young Earth Creationists presume what they assume to prove, that the Creation days of Genesis are 24-hour days. Rather than reading these verses as a commandment to rest, Young Earth Creationists train their disciples to read them as an instruction concerning the *duration* of the days of Creation! If one reads these Scriptures with that in mind, he or she may forget the purpose of the Sabbath rest as do those who wrongly suppose they are obeying this commandment simply by recognizing particular days of the week.

Regardless the Young Earth Creationist conceit that 24-hour Creation days are literal while Creation days of a longer duration are not, this is not the case. No text in the Bible literally teaches these Creation days as being of 24-hour duration. Rather, 24-hour Creation Day is a scientific reading imposed on the text that became popular following Archbishop Ussher's chronology of the nations and due to Isaac Newton's Absolute Time, thereby influencing scientists in the centuries that followed. So long has our world been taught that the duration of the Creation Days are 24-hour days that most will read the Scriptures this way. I have never read them that way because the duration of the Creation Days is nowhere given in this passage. Nor is a 24-hour duration of the Creation days mentioned anywhere in the first chapter of Genesis. Translators are supposed to translate as best they can the exact meaning of the original Hebrew text. But I have never found a translation of the Bible that translates these as 24-hour days. Thus, a 24-hour day reading is *eisegesis,* by which is meant reading the text by some imposed external light.

An important difference in our 24-hour day work week and God's six days of Creation is that our work renews following the Sabbath. This is not the case for the six Creation days which ended God's original Creation regardless of what evolutionists may suppose. Thus, man will always be the last and highest of God's earthly creatures, the one made in God's own image. In line with this, Hebrews 4:3-4 tells us that God has been resting from his work of his first Creation to this day. If the seventh day of Creation is at least six thousand years, so might the other days be of long duration.

The Scriptures do however literally teach

With the Lord a day is like a thousand years, and a thousand years are like a day. [Psalm 90:4; 2Peter 3:8]

The days of Creation were not our days but the days of the Creator who is the one who experienced these days. God was not on the clock when he was creating the world! 'Evening and morning,' the beginning and end of these Creation days, divide God's acts of creation specifically because these were not ordinary days! Long eras regarded as days also have evenings and mornings. Thus we speak of certain events as marking the dawning of a new day, while other things occur during the twilight of a previous age. The sun which God gave us to mark our 24-hour solar days was not even created until the fourth day of Creation. God does not live in, nor is he subject to time as Young Earth Creationists and Open Theists wrongly suppose. Rather, God is sovereign over time. God transcends time as he creates time by his acts of Creation.

The Hebrew word *yom* is used in at least four different meanings in the first two chapters of Genesis, as *light* (Gen 1:5), as the *days* of Creation (Gen1), as *the entire period of Creation* (Gen 2:4), and *when Adam would die* if he ate of the tree of knowledge of good and evil (Gen 2:17). Adam did not die within 24-hours of eating the forbidden fruit.

While acknowledging that *yom* is often used in the Scriptures for indefinite periods of time, Creationists claim that a numbered sequence of days in the Scriptures always refer to a 24-hour day. Do they believe that Hosea 6:2 refers to 24-hour days?

After two days he will revive us;
on the third day he will restore us,
that we may live in his presence. [Hosea 6:2]

There are verses in Ezekiel and Daniel that equate days with years regarding judgment and prophecy including a sabbath rest to the land lasting for one year. But whatever be the case about how days are used elsewhere in the Scriptures, the origin as well as the end of the world are exceptions to many things. Made-up rules so as to accord with some agenda are no argument for 24-hour Creation Days.

THE HISTORICAL CHRISTIAN UNDERSTANDING OF THE SIX DAYS OF CREATION

Interestingly, we hear both from Young Earth Creationists and those who wish to disparage the Bible that Christians have always read the Creation days as being of 24-hour duration. In support of their claim, Young Earth Creationists fondly quote the liberal Old Testament scholar James Barr (1912-2006) who delighted in using this interpretation to attack the Bible!

Until the rise of modern science, the duration of the days of Genesis was not a subject of great debate. A few church fathers did comment on the *hexameron,* a Greek term referring to the six-day creation which ended with the creation of Adam. They mostly used it against the eternalism of the world as taught by Greek philosophers such as Aristotle. But an ordinary day did not seem long enough for all the kinds of animals to be brought before Adam so that he could name them, then be put to sleep for the creation of Eve. Thus, some church fathers pointed out that the sun which measures ordinary days was not created until the fourth creation day. On the other hand, the philosophically minded Augustine of Hippo taught that the entire six-day Creation was instantaneous. Challenging Augustine, Martin Luther famously called 'a spade a spade' by declaring these ordinary days. The church fathers however were far more interested in what might be the duration and end of the world which they dated from Adam.

As I mentioned above, 24-hour duration of the Creation days were indeed presumed from the time of Archbishop Ussher's famous *Chronical of the Nations* endorsed by Isaac Newton who taught his proposed Absolute Time and Space as the very 'sensorium of God,' as if God himself is a physical being! As a cosmos far beyond the sun and moon were hardly considered until the invention of the telescope, few supposed that the earth might be very old until the scientific study of extinct fossils. Young Earth Creationists are troubled by the *Deep Time* of the geologists. Should not they have the same concern about the *Deep Space* of the astronomers?

For so long as he allows it to exist, the vast physical universe is but dust at God's feet and the vast era of creation of the heavens and the earth is but a moment, rather, but six days to God. No matter how old the physical world, physical time is of no consequence for the eternal God who so looked forward to creating man in his own image that what to us is a vast duration was but a short time to him. Would one rather that God had done nothing during the vast ages prior to his Creation of man? The earth is God's footstool, the place from where he rules the physical universe.[2] This planet is where he placed man. This 'pale blue dot' is the "Garden of Eden" of the physical universe. All that we see with our physical senses or measure with our physical instruments and clocks are but a shadow of now invisible Heaven, which is far greater, more interesting, more concrete and enduring place than the physical heavens that God spreads out like a tent.

We should however understand why the Bible does not tell us just how large and how old the physical universe. This is because the Bible is written for all ages and cultures and can thus be translated in terms that all ages and cultures can understand. This is why the Creation

---

[2] Isaiah 66:1, Matthew 5:35, Acts 7:49

account is written for the world that we see with our physical eyes and observe without extensive scientific research. But uniquely among the world's writings, the Bible reveals man to have been the last of God's Creation, which became essential for understanding an earth that is far older than man!

When the Scriptures refer to the most ancient times, they do not mention man, but rather the mountains, which is precisely where geologists discovered the great antiquity of the earth.

Your father's blessings are greater than the blessings of the ancient mountains, than the bounty of the age-old hills. Let all these rest on the head of Joseph, on the brow of the prince among his brothers. [Genesis 49:26]

He stood, and shook the earth; he looked, and made the nations tremble. The ancient mountains crumbled, and the age-old hills collapsed— but he marches on forever. [Habakkuk 3:6]

In Chapter 22, I will discuss the nineteenth century dispute over the age of the earth between the professional geologists and those called *scriptural geologists* who insisted on studying geology in the light of the Bible, as I also do. This earlier dispute however mostly concerned whether Noah's Flood was an anthropologically universal but geologically local Flood versus there having been a worldwide Flood. All the scriptural geologists believed in a worldwide Flood while most though not all of them taught a young earth. In their day a young earth had long been the view of the world's top scientists. By contrast, today's Young Earth Creationists anchor their teachings, not according to the worldwide Flood which they do teach, but for reasons that I explain below, so as to teach 24-hour Creation days.

In the first half of the twentieth century, the 'ruin and restoration' old earth view, which I will explain in Chapter 62, was the most popular view of Fundamentalists and taught in the very popular Scofield Reference Bible used by many Fundamentalists. The first anti-evolutionists such as William Jennings Bryan, the prosecutor at the Scopes Trial against the teaching of evolution, believed that each Creation Day lasted for one thousand years. So did many of the most conservative Christians, included those that wrote the pamphlets from which the word *Fundamentalist* derives. A young earth view has not always been the view of the most conservative and pious Christians, even in my childhood during the fifties and sixties!

However, Seventh Day Adventists Ellen G. White (1827-1915) and George McCready Price (1870-1963) believed that Noah's Flood explained fossil fuels like coal, oil, and gas.[3] This was due to a vision in which the Lord carried Prophetess White back in time to witness the creation as six 24-hour days. Her vision confirmed that the ten commandments of Moses including the seventh day Sabbath still apply to Christians as Seventh Day Adventists insist. This explains why some of defenses of the young earth view, as the case of vegetarian carnivores, reflect the vegetarian teachings of Ellen G. White.

These Neo-Judaizing teachings of the Adventists however were a hindrance for the acceptance of these teachings by evangelicals until the 1961 publication of *The Genesis Flood* by John C Whitcomb and Henry Morris. Though aiming to challenge local Flood arguments, the authors scarcely mentioned their dependence on the geology of Price. Since the publication of Whitcomb and Morris's book, which was popularized in Christian and home schools, most evangelicals have become Young Earth Creationists. This however points to a scientific defense of a worldwide Flood, which the Bible clearly teaches, as being the chief reason that evangelicals such as Whitcomb and Morris became Young Earth Creationists. Though they taught 24-hour Creation days, Whitcomb and Morris avoided the Seventh Day Adventists theological arguments.

Morris and Whitcomb however pushed back the date of the Flood by several thousand years to accommodate the massive archaeological evidence of early man, supposing that the Flood took place when mankind was in the Stone Age. This, in turn, pushed back the beginning of man to perhaps 10,000 years. Whitcomb and Morris recognized the problem in the archaeological evidence which as I will explain below comes from misunderstanding the geological evidence of Noah's Flood. What was then called *Flood Science* was changed in the 1980s to *Creation Science* in an unsuccessful attempt to get this theory taught in America's public schools. The popular term *Creationist* derives from this change. Though I commend Young Earth Creationists for holding to the truth of a worldwide Flood, we now have a more biblical and scientifically sound defense of a worldwide Flood than the provisional one provided by the Seventh Day Adventists.

---

[3] Ronald Numbers, *The Creationists: the Evolution of Scientific Creationism* (1992, University of California Press)

The Simplicity of Truth

Most Creationists may be unfamiliar with the scientific teachings intended to support their view. But If you have studied the Young Earth Creationist explanation concerning the Flood, you will have found many of them to be highly fanciful and no less 'man-made' theory than the ones they oppose. Their explanations of how humans managed to outrun the rising waters of the Flood so as not to get buried and petrified as did the animals can be hilarious. To mention again, though I point out these biblical and scientific contradictions, it is not my aim to ridicule this theory which for many years has been the only explanation of a worldwide Flood available to believers. I will point to many places where the Ice Age narrative taught by mainstream science is no less hilarious. Young Earth Creationists were right in teaching a worldwide Flood and mainstream science and most of my Old Earth Creationist brethren have been wrong in denying it. Moreover, some of what the Creationists have taught about Noah's Ark and how current day humans and land animals originate from the Ark only less than five thousand years ago remain valuable contributions to science.

My search for evidence of the Flood as narrated in this book is due to Young Earth Creationists having pointed out that Jesus presumed a worldwide Flood. While I am scientifically closer to Progressive Creationists who can also point to the astronomical, geological, and faunal evidence of Creation, I may be spiritually closer to most Young Earth Creationists who do not hold mainstream scientific views with such unskeptical reverence. There are however sincere Christians on all sides of the understanding of origins.

What makes any scientific narrative valuable is how simply it explains the evidence without requiring numerous *ad hoc* explanations to explain away so much evidence that appear to conflict with the narrative. We see this problem in the case of evolution and man-made climate change. A true narrative of the past will also be able to predict things that other narratives cannot, as in the above example of how the civilizations ended by the Flood are differently buried. This is also why the founders of scientific geology proposed what they believed was perhaps a worldwide Flood to explain a variety of evidence that could not be otherwise understood. Unfortunately, by replacing the Diluvium Flood with a supposed frozen flood called the Ice Age, mainstream science established the unsound foundations of the utterly confused and muddled understanding of the past that is currently being taught.

4

DATING THE PAST

If you are acquainted with the scientific history of man's past, you may have supposed that the great antiquity of man has been confirmed by scientific dating such as radiocarbon. Radiocarbon can date organic remains that are as old as perhaps 50,000 years. That it can be an accurate means of dating is confirmed by paleographic-determined calendrical dating of the Dead Sea Scrolls. Due to the rare conditions of having been preserved in dry caves that are below sea level, thus more protected from damaging ultra-violet rays, these scrolls of leather and papyrus have survived for more than two thousand years. The Dead Sea Scroll dating scientist Dr. Stephen Pfann has also been involved with this discovery because he also helped identify the sources of clay of the ceramics of ancient migrations which begin near Mount Ararat.

We tend to trust technologies which we believe give objective results, especially if we have learned what appear to be the straightforward physical methods by which these technologies obtain their results. But in the case of prehistoric dating, this most basic form of scientific dating is hardly straightforward. This is because, aside from special conditions, organic artifacts rarely last for thousands of years. How many wooden houses, furniture, leather goods, or precious books do you know that are older than one thousand years? Large bones and antlers last much longer, especially if buried in dry places. The remains of the Ark have been preserved due to having been constantly frozen. But the difficulty of dating things from many thousands of years ago as the case of most prehistoric sites is knowing just what is being dating.

Archaeology depends on two kinds of dating, *relative* and *absolute*. Relative dating is determined by archaeological context, which is essential for knowing just what it is that we are dating. Relative dating is presumed to follow the archaeological deposits of successive ages, the ruins of later periods being found above those of earlier periods as we do see in the case of historical periods. In the lands of the Bible, Roman era artifacts lie above those of Hellenistic times which lie above those of the Persian period, which lie above Iron Age deposits of the kingdoms of ancient Israel. But this is not the case for the prehistoric periods where relative dating is determined by archaeological type styles such as a particular design of stone tools or clay pottery. Such type styles currently define certain ages as I will explain below, but due to not being found in successive strata, type styles that define successive archaeological periods may have existed at the same time. In fact, most did co-exist in the antediluvian world.

Fig 5 - Walter Libby (right), discoverer of radiocarbon dating

Radiocarbon is a form of absolute dating. It tells us the age of the organic samples that we use for dating. But the relative dating of prehistory long preceded the advent of absolute dating by radiocarbon, which was invented only in the late forties (see Fig 5).

Too little attention has been given to the relative dating of our scientific prehistory upon which the scientific narrative depends. As I will explain in Part IV below, the relative chronology used for prehistoric archaeology is based on a combination of a mistaken geological chronology and a mistaken artifact-based chronology. The mistaken geological chronology is due to (1) not recognizing supposed Ice Age deposits as being the result of Noah's Flood and (2) failing to recognize that these deposits also cover the prehistoric burial mounds that are currently interpreted as elaborate religious burials. In fact, these "prehistoric" peoples were those who drowned in Noah's Flood!

The chronology now used for prehistory is not based on archaeological strata as might be supposed, but on a deeply racist understanding of living tribes and nations. These vile racist views developed from replacing a Christian/Heathen understanding of the nations with white/colored racism to explain the progress of European civilization apart from Christianity. Tribes using stone tools as late as the nineteenth century were accordingly called *primitives,* those incapable of progress apart from the help of white Europeans. Today, racism aimed at contemporary peoples is abhorred, but the same does not seem the case for the ancestors of living peoples which lie at the foundations of the current scientific prehistory.

Radiocarbon dating has pushed back the era of prehistoric sites, but interestingly, it scarcely changed the prehistoric relative dating that existed prior to radiocarbon dating. The chief reason is that artifacts are rarely used for the radiocarbon dating of prehistoric sites that are as old as about four thousand years. Instead, what is used for dating these sites are the typologies based on an unsound relative dating together with radiocarbon dating of amorphous charcoal and organic deposits from the ancient earth which may have existed long before the humans who occupied these sites. There may be soil clinging to the foundation of the house where you live that contain tiny bits of charcoal or decayed remains that are as old as tens of thousands of years. If we blindly trust radiocarbon dating, you may find yourself living in the world's oldest dwelling. You know better but neither do you have an incentive to make such a claim.

As the case of almost everything today, archaeology depends on funding. Funding however depends on the excitement of digging and reporting the oldest this or that kind of site. This explains why the dates of prehistoric sites keep getting pushed to ever more ancient times but are almost never corrected to more recent dates. Thus, the prehistoric world becomes ever more mysterious and hopelessly muddled. Many of the radiocarbon dates may be useful but knowing just which of these dates are valid requires understanding what has been dated. I will explain more about dating below, including the very interesting story of how the scientific history of man's past came about but, more important, how the deeply muddled and confusing scientific view of man's past can be simplified and clarified by seriously using the light of the Scriptures.

5

NOAH'S ARK AND FLOOD, A SUPERNATURAL AND JUST EVENT!

I will explain in considerable detail the natural processes involved in Noah's Ark and Flood. The miracles of the Bible are not as some teach, a suspension of the laws of nature. Otherwise, an Ark might not have been necessary to save people from drowning in the Flood. Nor was the Bible written to people who did not know the ordinary laws of nature as seems the case of pagan myths. Such myths concern fantastic worlds where strange powers appear ordinary. Rather, the miracles of the Bible took place in a world like the one we experience. This is why the miracles of the Bible so impressed those who experienced them. The supernatural miracles recorded in the Bible are due to the special presence and purposes of the all-good, all-wise, and all-powerful God of the Bible.

Nor is the matter of Noah's Ark and Flood just an ancient wonder that happens to be recorded in the Bible. It has an essential role in the Bible's greater theme of salvation and judgment and for explaining just who is the God of the Bible. Aside from its place in the account of origins in the early chapters of Genesis, Noah's Flood is mentioned in the Old Testament book of Isaiah which differentiates the God who created the world from the pagan gods who are part of the world. Jesus and his Apostles use this event to explain the Great Day of salvation and judgment that accompanies the Second Coming.

Those inescapable miracles

I will be explaining how the men of the Enlightenment rejected the miracles and divine revelation of the Bible and how the evidence of Noah's Flood was reimagined so as to somehow explain away this miracle. Nonetheless, the world of science has been unable to escape miracles. There is no *scientific* way to explain such singularities as the

Fig 6 – The Dove Sent Forth from the Ark

beginning of the universe, the organization of the cosmos so as to provide a place suitable for life, or the beginnings of life in this favored place of our universe. If one believes that life began by chance, that is the greatest of miracles but one that no reasonable person should believe.

A *singularity* is but another word for a miracle. Scientists must accept certain singularities which they cannot scientifically explain so as to begin their work. When historicity becomes the best way to explain the larger context of some matter, it deserves to be believed even if a miracle. Just as scientists learn to accept the miracle of the Creation of the world, the exceptional conditions of our privileged planet and the miracle of life's beginnings, I will show how the historicity of Noah's Flood and Ark is the only good way to understand the recent history of the earth and the earliest evidence of man.

Noah's Ark and Flood is arguably the most astounding nature miracle recorded in the Bible, not excluding Creation itself. Of all the miracles in the Bible, the great skeptic Voltaire found Noah's Flood to be the most difficult to believe. Of course, neither did Voltaire wish to believe, so much so that he claimed that seashells found in the mountains that many scientists of his day attributed to Noah's Flood were carried there by pilgrims.

For sceptics seeking truth, on this much we can agree. Due to the ordinary laws of nature there ought to be plentiful evidence of something that impacted every human being, every artifact made by humans, every animal, every plant, and every parcel of land upon the face of the earth. The evidence should be massive, detailed, and clear. Though this evidence has been hidden from modern eyes by the current scientific narrative of the past, the evidence is more plentiful than anyone could have imagined.

The miracle of Noah's Ark

While the natural processes were involved with this event, so was the supernatural power of God. I have in mind such things as the drawing to the Ark of a pair of all the kinds of animals which became the ancestors of every living species. No less miraculous was the protection and provision of food for these animals as they migrated to the places assigned them after the Flood, which were not always the same places their ancestors had lived before the Flood! Only the God who knows when every sparrow falls[4] could have known how to restore the earth and save these species during this time of such tiny seed stock populations. Apart from God's supernatural protection, all the species that now exist would have gone extinct following the Flood.

However, the Flood has not been the only time when God has supernaturally intervened to protect his creatures. The Scriptures teach that God regularly does this. Otherwise, neither we nor any of God's creatures would be long upon this earth. Consider the constant dangers to the earth from asteroids and comets. Life on earth will always need God's protection and management. Throughout the Scriptures, we are reminded that God not only supernaturally regulates the climate, he also determines the weather. Thus, there will always be limitations in our ability to mathematically predict climate, weather, and earthquakes. This is why man's ability to predict the future has such great limitations.

God not only provides for his creatures through his providential processes that scientists discover and study, but he also does this in response to the needs or the correction of his creatures. Though God intervenes on a regular basis to manage our climate and weather not only for our benefit but also to bring us closer to him through correction and sorrow, we do not call these 'acts of God' *miracles*. This is because they are not so extraordinary as the case of Noah's Ark and Flood. In Chapter 60, I will explain the natural cause of a 12-year cycle in sunspots and weather that scientists have yet to recognize. Even so, due to responding to the needs or correction of his creatures, which are otherwise free to disobey him, the climate or weather will always be something that scientists can never completely predict however soon they can observe these changes and from that, hope to predict the weather or earthquakes for days, weeks, or perhaps years. God of course knows and controls the future and can show them to his

---

[4] Matthew 10:29

prophets as in the case of the patriarch Joseph, the prophet Elijah, and the New Testament prophet Agabus.

THE JUSTICE OF THE FLOOD

Skeptics complain of the justice of the God of the Bible for sending the Flood. They forget or, more likely, do not believe that the world before the Flood was as wicked and corrupt as the Bible says. In Chapter 8 below, I will show that this was indeed the case. Nor do they believe the clear teaching of the Scriptures that ultimate justice will follow our death, that Jesus Christ will one day judge all the living and the dead.[5] Jesus preached his salvation even to those who died in the Flood.[6] The same God who made us also decides how and when each of his creatures die however much our behavior impacts his judgment.

We should be thankful that God is kinder than us. So terrible was the world before the Flood with its brutal slavery, terror, and violence, that the Flood was a mercy killing. As one who has on several occasions nearly experienced drowning, I remember those occasions as being peaceful events. My thoughts were completely on the foolish things that I had done with my life, including putting myself in danger of drowning.

CONTEMPORARY EVIDENCE OF THE BIBLE'S GREATEST NATURE MIRACLE

Many would like to witness a miracle, but miracles are rare. Otherwise, they would not be miracles. There are also counterfeit miracles as in the case of magic or perhaps things done from preternatural powers. However, God used authentic miracles to confirm his witness and presence as in the case of his Prophets and Apostles.

Though some today testify to having witness a miracle, the last of the great public and history-changing miracles that established the foundation of the Christian faith occurred almost two thousand years ago. The more distant from some great event and the more time that passes, the more skeptical the world becomes. A world that no longer recognizes contemporary miracles has even less reason for believing a great miracle from so long ago. This was the argument of the great skeptic David Hume for rejecting the miracles of the Bible. But if Noah's

---

[5] John 5:22,27

[6] 1 Peter 3:19-20

Ark has been found, the world no longer has a good reason for rejecting any of the miracles of the Bible.

For sure, Christians should believe the historicity of Noah's Ark and Flood because the God who created the world and can save people from death can hardly do so aside from possessing such power and wisdom as was necessary for Noah's Ark and Flood. This may be the reason that God allowed this episode in the history of salvation however much it grieved him to have to send the Flood. Thus, Jesus referred to this same event to compare the great destruction of the world but likewise the miraculous salvation of the faithful with his Second Coming.

Aside from the truth concerning Noah's Ark and Flood, everyone ought to know there is no way that we could have come to exist or continue to exist for another day apart from the goodness of the one and only God of the Bible. Even if some might have somehow come to exist, why should you and I be among those so favored? Such miracles require us to humble ourselves to one who is wiser, more powerful, more righteous and greater than ourselves. It is arrogant for us to suppose ourselves God's equal, that he should submit himself to our demands for evidence. Those who do so raise themselves to equality with God as did Satan for whom he created the Lake of Fire mentioned in the book of Revelation. But he gives grace to those who humble themselves as little children who so love the story of Noah's Ark.

Those who do so humble themselves are far more likely to experience a miracle. But God is merciful even to those who are not righteous or humble. Though he showed his resurrected body only to his disciples, Jesus offered his empty tomb as evidence of his words to a 'wicked and adulterous generation 'who were undeserving of getting to see any miracle.[7] His preservation of Noah's Ark and its revelation at just this time when the world is in such turmoil is perhaps no less a miraculous public witness to his promised Second Coming. Prepare yourself to discover that the current historical sciences supposed as the product of the world's greatest minds is not only false but hilariously false. But also prepare to discover that Noah's Ark which today is believed mostly by little children is profoundly true. Such is the great and wonderful God of the Bible!

---

[7] Matthew 12:39

Part II

The Archaeological Evidence of the Flood

6

A Systematic Search for the Archaeological Evidence of the Flood

Young Earth Creationists focus their defense of Noah's Flood on geology. But archaeology concerns the material remains of the human past which are more recent and thus better preserved from the ravages of time than the far more complex and difficult to understand geological past. Thus, my investigation began with an objective methodological search for what should be manifest archaeological evidence of a historical worldwide Flood. While I knew that both mainstream and Young Earth Creation scientists deny there exists such evidence, my scientific experience teaches me to be a skeptic. I rarely find the teachings of specialists to be as sound as claimed. My newly found faith in Jesus required me to make this search because I do investigate what appear serious claims against the truth of the Bible. I had at least to satisfy myself that denials of Noah's Flood truly rested on firm foundations.

I also knew from scientific experience that one can directly look at the evidence for which one seeks and not recognize it, especially if it concerns something unique. While we familiar with the evidence of ordinary floods, we are not accustomed to the kind of Flood that could float the Ark to the top of Mount Ararat. I aimed to find out whether such evidence exists and, if so, what might be its geographical extent. The methodology I employed was to establish criteria that should serve as manifest archaeological evidence for Noah's Flood as taught in the Bible. For all I knew, these criteria could have disproven the historicity of Noah's Flood! Whatever the case, I wanted to know the truth. Thus, I chose to investigate whether there existed evidence of

(1) a sudden disappearance of the earth's population in the third millennium before Christ,

(2) which should be followed by a secondary worldwide dispersion of people from the lands of the Bible,

(3) with a clear break in material culture every place where appears this secondary dispersion,

(4) and evidence that the sudden ending of the world's population was due to drowning.

I did not initially look for evidence of drowning until I recognized it in the archaeological sites of the world's first civilizations. But those would include most prehistoric sites on every continent.

After several years of unfruitful searching in texts of ancient history, I gave up searching. But as a result of my search, I became interested in the archaeological history of the lands of the Bible, the prehistoric chronology of these lands being listed in Table I below. The Paleolithic (Old Stone Age) eras are the remains of those who lived in these lands during a supposed Ice Age. The Epipaleolithic refers to the era near the end of the Ice Age. The Neolithic (New Stone Age) eras which mark the first permanent settlements immediately follow the Ice Age before the use of copper which marks the Chalcolithic era. The Early Bronze Age marks the high-water mark (pun intended) of mankind's earliest civilization. It was a materially advanced civilization!

Table I

Prehistoric Chronology

Middle Paleolithic	300,000 - 30,000 BP
Upper Paleolithic	30,000 - 12,000 BP
Epipaleolithic	12,000 BP - 8,300 BC
Pre-Pottery Neolithic	8,300 - 5,500 BC
Pottery Neolithic	5,500 - 4,500 BC
Chalcolithic	4,500 - 3,300 BC
Early Bronze Age	3,300 - 2,300 BC
Middle Bronze Age	2,300 - 1,550 BC
Late Bronze Age	1,550 - 1,200 BC
Iron Age	1,200 - 586 BC

BP = years Before Present

To my great surprise, I happened upon the very kind of evidence for which I had been searching. At the end of the Early Bronze Age, the population of the lands of the Bible disappear! I did not make this discovery. Archaeologists have already found it! To quote a prominent Israeli archaeologist:

A phenomenon so central in Palestinian history as that of the formation of urban society, with all its consequences for the material and spiritual culture of the country dissolves before our very eyes. [Ammon Ben-Tor, *The Archaeology of Ancient Israel,* (Yale University Press,1994), p 123].

I suspected that the solvent for the Early Bronze Age cities may have been water. Indeed, the Early Bronze Age ended about 2300BC, which is very near the date of the Flood as can be calculated from the Bible! Here was evidence of my first criteria. The clear break in material culture between the Early Bronze Age and the Middle Bronze Age that follows was evidence for my third criterion for the archaeological evidence of Noah's Flood. I soon discovered that this break which follows the Early Bronze Age exists across the earth.

We can observe this break between the Early Bronze (EB) and Middle Bronze (MB) Ages in archaeologist Kathlyn Kenyon's famous drawing of the strata of ancient city of Jericho. She even found what she labels fine *water-laid silt* lying just above the Early Bronze strata (see drawing below).

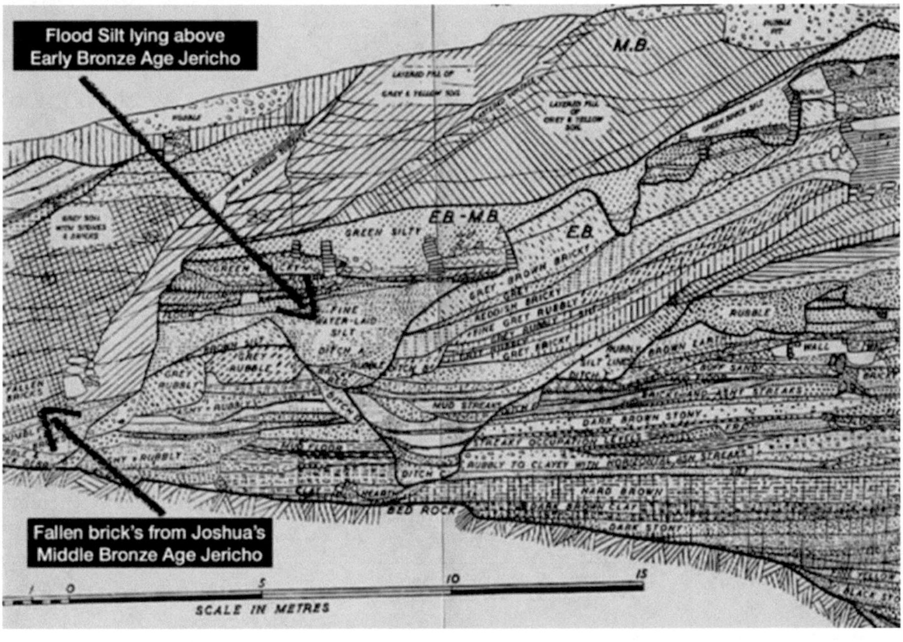

Fig 7 - Kathyleen Kenyon's Sketch of Jericho

Contrary to claims made against the historicity of the Flood, "the world's oldest cities" were not continuously occupied for 10,000 years. The supposed continuity of occupation of Jericho, which has long been touted as the world's oldest city, has long been used as evidence against the Flood. But Jericho was completely destroyed on at least two occasions (1) the end of the Early Bronze city by the Flood and (2) the city's destruction by Joshua. The bricks from the fallen walls of the Middle Bronze Age city occupied by the Canaanites are evidence of the latter. In the ruins of the ancient city of Jericho is evidence of two of the greatest miracles mentioned in the Bible!

Interestingly, Dame Kenyon writes about the great erosion that took place all over Palestine at just this time. This is something that exists the world over but is largely ignored by archaeologists, though not by geologists who discover a great number of "local" Floods. Among these are the Black Sea Flood, the so-called Missoula floods in the American Northwest, the Grand Canyon Flood, the Flood in Tibet and other places in the world supposed as having been caused by the breaking of water barriers or natural dams. All these are local manifestations of what was in fact a single worldwide Flood.

It seems that no one had previously connected this disappearance of people immediately following the Early Bronze Age with the Flood though it happened at the very time of the Flood as given in the Bible! It is however rare for an archaeologist to make a claim that would conflict with everything that current science believes about the ancient world. It is also a difficult thing to do because everything else in the historical sciences have been interpreted according to the narrative of a supposed Ice Age that I am challenging with the simple narrative of the origins of man as given in the Bible.

7

THE FIRST CIVILIZATION AFTER THE FLOOD

Who are the people that repopulate the lands of the Bible during the Middle Bronze Age and from where do they come? The first civilization to appear in the lands of the Bible after this hiatus in population following the Early Bronze Age are the Canaanite city states, also mentioned in the Bible as the first civilization in this land following the Flood. These Canaanites settlements, which include the famous Ebla in Syria, mark the Middle Bronze Age in the *Levant*, which is a reference to the combined lands of Syria, Lebanon, Israel, and Jordan. While Egypt's Old Kingdom that followed the Flood is presumed to date to the Early Bronze Age, the many scarabs of Old Kingdom Pharaohs found in the Levant are all from Middle Bronze Age sites. These Middle Bronze Age migrations correspond to the Sumerian and Akkadian periods in Mesopotamia, the Sumerian civilization being the oldest known historical people who possessed the only language that transcended this great interruption. Though becoming the greatest civilization of the Ancient Near East following the Flood, the archaeological remains of the Akkadian civilization are tiny relative to the remains of the deeply buried Early Bronze cities ended by the Flood.

The Middle Bronze Age begins with migrations as shown on the map on the next page (Fig. 8) published by Ammon Ben Tor, the Israeli archaeologist quoted in the last chapter. As indicated by the arrows, the origins of this migration spreading into ancient Canaan should be some place close to Mount Ararat! Depending on the countries where it spreads, it is variously called *Transcaucasian, Kura-Araxes,* or *Khirbet Kerak*. Transcaucasia is the region straddling the South Caucasus mountains which, as you can see on this map, run the Kura and Araxes Rivers that are just north of Mount Ararat. Archaeologists who study this culture tell us:

Kura-Araxes traditions apparently emerged during a rapid burst of cultural evolution, the early stages of which remain undocumented or unrecognized. [Peter Glumac and David Anthony, *'Culture and Environment in the Prehistoric Caucasus: The Neolithic through the Early Bronze Age,'* published in *Chronologies in Old World Archaeology, Third Edition, Vol I* (1992, University of Chicago Press), edited by Robert W. Ehrich, p.203]

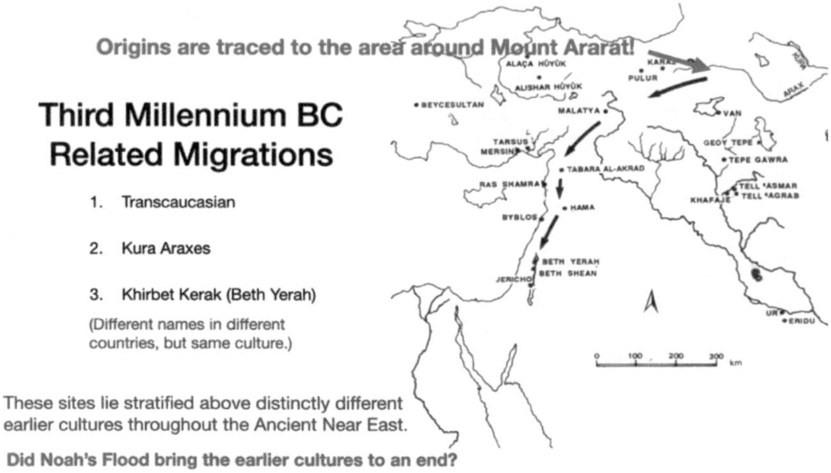

Fig 8 - Migrations from Mount Ararat

An archaeological *culture* consists of a family of related artifacts presumed to have been produced by a distinct people group. We may soon be able to document the amazing early stages of this culture should it be found to have begun in the ancient wooden structure that I visited near the top of Mount Ararat, precisely what led me to defend the Mount Ararat Discovery.

Khirbet Kerak, the third term used for this culture, is the Palestinian name for the ancient ruins found near where the Sea of Galilee empties into the Jordan River. It is the place where the eminent biblical archaeologist William F. Albright first identified this culture, which also defines the Early Bronze III (EBIII) era in the Ancient Near East. Notwithstanding how Albright named it, the Early Bronze III era is the true beginning of the Middle Bronze Age which, as we see on the map, begins near Mount Ararat and migrates along the path shown by arrows into Palestine.

Because the Early Bronze Age cities had just came to an end, Albright supposed these new people were invaders. But in contrast to the violence besetting the Early Bronze Age cities, these newcomers were peaceful. Such a peaceful culture could not have spread during a time when the violent civilizations of the thick-walled Early Bronze Age cities existed. Those spreading from the mountains of Ararat did often settle in or near the silt covered ruins of the Early Bronze Age cities which were locations central to travel or else had good water sources, rich land, or defensible sites.

The EBIII culture is identified by its distinctive red-black burnished ceramics which obtain their red coloration due to being baked in an open fire. This first culture after the Flood which we might suppose as that of Noah's family began in the ancient mountains of Ararat which the Bible mentions as the landing place of Noah's Ark. Most interesting, the settlements of this first culture are found in the highlands as if the lower elevations are still covered by water![8] The thin-walled, key-hole shaped building foundations associated with these sites are suitable either for tents or wattle-and-daub housing. The latter may have developed by turning tents into permanent residences as we might expect of migrating populations. Fireplaces are found in the center of these houses where likely was made the red-black burnished pottery.

Dr. Stephen Pfann, the Dead Sea Scrolls dating scientist who I mentioned in Chapter 4, pioneered the use of neutron activation analysis using this same Khirbet Kerak pottery. Working with Charles Burney of the British Institute of Archaeology in Ankara, Pfann discovered that the pottery was both imported and locally made. That points to the pottery spreading by migrations rather than from trade, just what we should expect if this pottery was spread by families migrating from the Ark.[9]

There is another culture that extends over much of the same expansive region of the Ancient Near East as EBIII. It is associated with a city in ancient Mesopotamia called Uruk, but which the Bible refers to as Erech, one of the cities that Nimrod established on the plains of Shinar.[10] The name *Shinar* means 'the land of Sumer and Akkad.' The archaeological culture that spreads to or from this city which suddenly became the world's largest is called *Late Uruk*. Cylinder seals and bevel-rimmed bowls mark the spread of *Late Uruk*. Of the many

---

[8] Mitchel S. Rothman, 'KURA ARAXES CULTURE AREAS AND THE LATE 4th AND EARLY 3RD MILLENIA BC POTTERY FROM VELI SEVIN'S SURVEYS IN MALATYA AND ELAZIG, TUKEY,' *ORIGINI* XXXVI 2014, p. 39

[9] Stephen J. Pfann, "Khirbet Kerak," in *Anchor Bible Dictionary* edited by Noel David Freedman (Yale University Press, 1992)

[10] Genesis 10:10

cultures in the Ancient Near East, *Late Uruk* is by far the most widespread. These are just some of many reasons to suppose that the city of Uruk was also the site of the Tower of Babel.

As I will discuss in Chapter 49, the bevel-rimmed bowls and the wooden-mold used for making them appear to be among the artifacts reported to be found in the ancient wooden structure that I visited in 2014. So is the tree-of-life motif that is so widespread in the Mesopotamian culture of this era if not also some form of proto-Sumerian texts. The tree of life motif points to knowledge of the biblical account of the Garden of Eden. If confirmed, this will transform our understanding of ancient history.

Late Uruk is currently dated to the Chalcolithic era which is earlier than the EBIII. This is because it was presumed to have immediately followed the earlier Uruk phases, which however were ended by the Flood. The bevel-rimmed bowls, which were produced from a wooden mold, may have also served as a standard measure for trading grains and other produce while the cylinder seals served as a mark of ownership. Likely, the cylinder seal was invented at Uruk where cuneiform writing was also invented.

Some of these writings tell us that the origin of the Sumerians (a name which derives from Noah's son Shem) was in the land of Aratta near the origins of the Tigris and Euphrates Rivers just south of Mount Ararat. Arrata was obviously the land of ancient Ararat where later existed the Kingdom of Urartu. These Late Uruk markers may be spreading with the Khirbet Kerak, Kura-Araxes, and Transcaucasian EBIII culture all of which mark the migrations of Noah's family that begin at the Ark that I visited near the top of Mount Ararat.

The material culture of archaeological sites of the Middle Bronze Age in the Ancient Near East has numerous characteristics that spread across the earth along the migration routes on the map from the end sheets of my 2011 book *The Archaeological Evidence* (map on next page). The intercontinental migrations as shown on the map will be unfamiliar to most students of archaeology. As I will explain in Chapter 39, this is due to the cessation of the study of migrations and cultural diffusion that has informed archaeology since the sixties. Though they study migrations from Africa, archaeologists no longer study worldwide migrations of the historical nations as shown on this map, reckoning that ancient agriculture and civilization developed independently in every land and continent. But this situation will soon change.

From the Middle Bronze Age of the Ancient Near East come the earliest historical cultures of every continent. The rich material culture carried by the migrants vary but they include spindle whorls and the heddle loom; waddle-and-daub huts and multi-generational houses with interior courtyards (*megarons*); the lost-wax form of bronze casting; the bow-and-arrow; bark paper and the hieroglyphic scripts used to write the earliest texts; spoked-wheel wagons; agriculture and animal husbandry; side lugs for tying and carrying pottery; giant urns for storing grain; gray, sand-tempered pottery, some with eggshell thin styles that appear to have developed from metallic predecessors and chevron, swastika, and other patterns created by stamping and incisions; cremations, and urn burials. It is also characterized by the practice of animal sacrifice, recognizing among other gods a great Spirit (i.e. invisible God) who lives in the heavens along with memories of the great Flood.

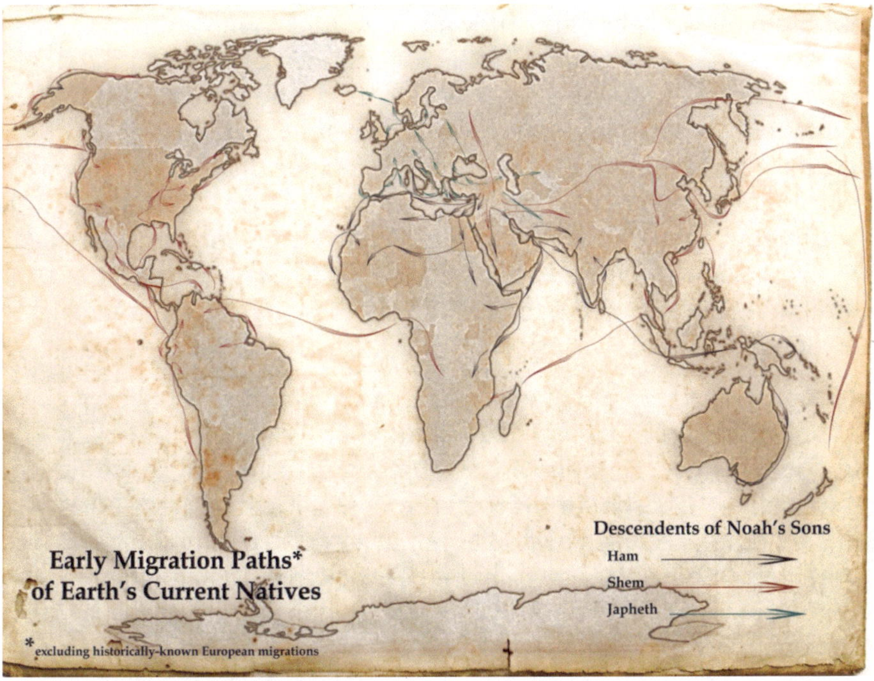

Fig 9 - Earliest Migrations of the World's Historical Nations

All this concerns the archaeology of those who survived the Flood, which met my second criterion for evidence of the Flood.

8

Civilizations Ended and Buried by the Deposits of Noah's Flood

The archaeological sites of those who lived in the antediluvian civilizations who perished in the Flood have been mistakenly interpreted as elaborate religious burials. They are ordinarily found under mounds of soil as the case of the so-called Indian mounds of America[11], the barrows of Great Britain, the kurgans of Russia, and similar tumuli found throughout the world. This is because these sites contain elaborate burials as we might expect from houses and cities drowned by the Flood. Most of these dwellings are covered by mounds of topsoil, sand, and gravel that were eroded by the Flood then settled on these sites as the Flood subsided. We can conclude these are not ordinary burials but the ordinary dwellings of those buried by the Flood by a number of observations:

1. There are almost no places associated with these burials where the people buried there might have lived. Some explain this lack of occupational sites by supposing they belonged to seasonal-visiting pastoralists such as shepherds or herders. But most of these burials are in the floors of houses owned either by farmers or city dwellers. The truth is these burials *were* the occupational sites of those buried in them. This includes those buried in the trenches around Stonehenge, the trenches being the long-decayed ruins of wooden shelters. The *mastabas* beside the Great Pyramids in Egypt were however the mudbrick ground-floors of the wooden dwellings of those who built the Pyramids. The Pyramids, likely built for protecting treasures in the decadent and violent era before the Flood, were later remodeled as tombs for kings.

---

[11] American burial mounds are earlier than the temple mounds of the historical natives.

2. Ordinary sites where people once lived have worn out or broken artifacts, the types of things that are ordinarily tossed into refuse. But these burial sites have unworn and even unbroken artifacts as would have belonged to occupational sites that were suddenly ended. These artifacts also include various industries with product stores in both complete and incomplete states. Many of these are explained as 'votive offerings.' Anything can be imagined as having been a "votive offering." There is no puzzling artifact that cannot be explained by some or another religious interpretation. These religious interpretations tell us about the interpreter but nothing about the religious beliefs of the people who lived there, even whether they had particular religious views concerning death. Absent relevant texts, religious interpretations should be abolished from the practice of archaeology.

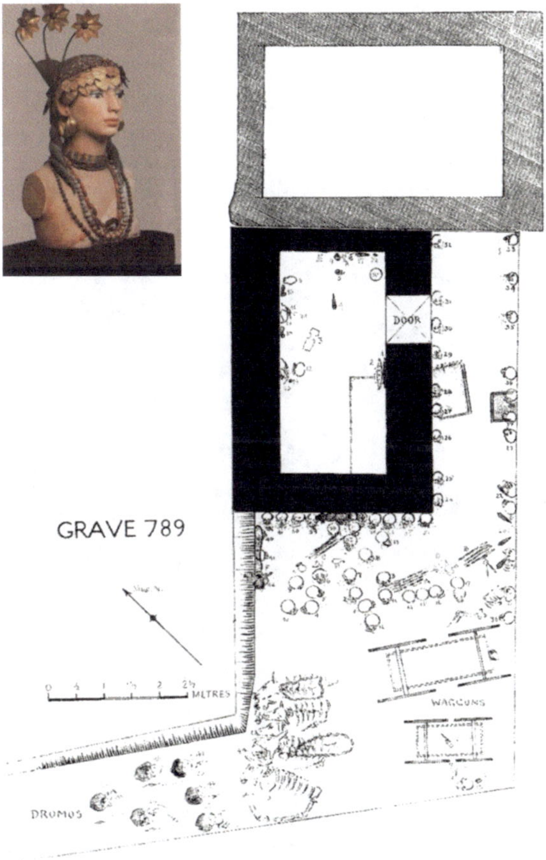

Fig 10 - Sir Leonard Woolley's drawing of ancient Ur grave where he found treasures as the gold headdress

3. Some of these graves are filled with great treasures of silver and gold. I got my first insight into the nature of these graves (Fig 10) by reading Sir Leonard Woolley's account of a supposed suttee funeral by which he explained why so many dancers and soldiers were buried at the same time with the King and Queen in the 'Royal Tombs of Ur.' Woolley complained of all the looting that must have taken place before he got his opportunity. The truth is, no one buries such great treasures in unguarded tombs unless, as in the case of ancient Pompeii, there be no one to witness the burial! I immediately recognized a party going on in the ancient city of Ur which

Noah's Flood brought to a sudden end.

Woolley's workers were already standing in the ruins of the Flood when they dug through 8 feet of water-laid mud, which in 1929 Woolley announced to the world as evidence of the "Flood," then assumed but a local one. This stream of heavy mud would have been the first and lowest layer of Flood deposits to flow into the city from erosion caused by the waters that eroded topsoil, clay, and sand from the Zagros mountains. Mud and topsoil covered the entire royal palace. Future settlers at the site of Ur would find what Wooley explained as a sacred place more suitable for a garbage dump. The covered ruins of Ur became but one of the numerous tels throughout the Near East, most of them initially created by Noah's Flood.

4. Ordinary burials are mostly graves of the elderly. But, as the case of the young dancers in Ur, these burials contain people of all ages and of apparent excellent health. Numerous infants are found buried in jars which may have served as their beds or wash tubs.

5. Every place across the prehistoric world, people are buried with weapons and tools in caves and fields that look like ordinary campsites, in tombs that look like ordinary houses, and in necropolises that look like ordinary cities. Some lie on matting and platforms that look like ordinary beds accompanied by household artifacts, many with personal weapons and jewelry.

6. Those in ordinary burials are placed in formally arranged positions. Not so those in these vast prehistoric burials, who are found in many casual positions as if by chance arrangement. Many of those buried in these sites are found in flexed positions (Fig 11), which is evidence of death by drowning. This happens due to arms and legs floating in water above the heavier torso. The corpses then settled in such

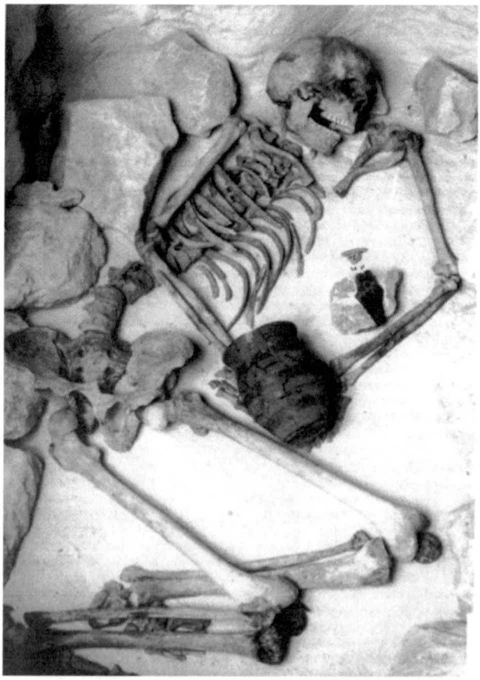

Fig 11 - Flexed burial of beer drinker with his dagger

flexed positions. Houses of great stones and timbers as well as cave dwellings kept bodies from floating away. But Otzi the Iceman, found in the nineties buried under a glacier in the Alps (see Fig 30, Chapter 47) was restrained by his quiver strap that caught on the rocks as he was being swept away by the waters of the Flood.

7. Vast numbers of lowland burials from prehistoric sites are found under mounds or tels. Archaeologists tell us that Early Bronze Age remains in river valleys and deltas are too deeply buried for easy excavation. By contrast, Early Bronze Age sites found in highlands are either slightly buried or else found entirely above ground. This is just what we should expect if the waters of the Flood eroded topsoil, sand, rocks, and clay from the mountains and highlands, sweeping this mixture of soils and rocks into the foothills and river valleys covering farms, villages, and cities. Many of these sites lie off the coasts of seas and oceans which contracted due to the expansion of the oceans resulting from water from the Flood.

8. The last thing I mention here as evidence of the world before the Flood comes from the light of the Bible, that is, it being a time of decadence and great violence.

So God said to Noah, "I am going to put an end to all people, for the earth is filled with violence because of them. I am surely going to destroy both them and the earth. [Genesis 6:13}

Archaeologists interpret the tiny vessels from this time as being used for drugs. There is also plentiful evidence of the use of great amounts of wine and beer. Excessive violence explains the numerous weapons,[12] injuries to skeletons, and the thick walls of these Bronze Age cities. To judge the level of violence, one need only search the Internet for 'violence, Early Bronze Age' or 'violence, Neolithic.' If you do this search, you will be shocked by the level of violence, the chief reason that God brought this bloody world to an end.

This ending of the world's ancient population confirmed the last of my four criteria for identifying the archaeological evidence of Noah's Flood. It was these Flood graves which convinced me beyond reasonable doubt that I had discovered the archaeological evidence of Noah's Flood. Archaeologists who read this chapter will never again be able to examine one of these prehistoric burials without thinking about the Flood.

---

[12] See beer beaker and dagger, Fig 11

9

How the Antediluvian World Got Expanded from Six Thousand to Hundreds of Thousands of Years

There does appear a problem with understanding these prehistoric sites as having been ended by the Flood. According to the current prehistoric chronology listed in Table I of Chapter 6 above, sites as I describe above as Flood burials extend for hundreds of thousands of years, from the Middle Paleolithic to the Early Bronze Age! The New World Paleolithic and Archaic sites are also Flood burials. It is however noteworthy, as I mentioned earlier, that prehistoric eras are defined by archaeological type styles rather than successive strata at continuously occupied sites. Most of these sites may have been contemporary!

This explains why the current prehistory of man jumps from Africa to the cavemen of Europe during the supposed Ice Age, then back to the Ancient Near East at the end of the Ice Age where begin agriculture and cities but still jumping from site to site for thousands of years. This however is due to a mostly imaginary prehistoric chronology based on typology which also attempts to correct the great hiatus following the end of antediluvian civilization. In truth, most sites assigned to everything from the Middle Paleolithic to the Early Bronze Age were more or less contemporaneous!

The lack of sites associated with the depleted Middle Bronze Age population that followed the Flood has resulted in some archaeologists pushing down the end of the Early Bronze Age from about 2,300 BC to 2,000 BC and pushing up the Late Bronze Age from about 1,400 BC to 1,550 BC.

Scientific archaeology got its beginnings in the early nineteenth century by those studying prehistoric sites in Scandinavia. Some of these ruins contained artifacts of bronze and iron, but others had only artifacts of stone. Christian Jurgensen Thomsen (1788-1865) decided the way to date these ruins from earliest to later would be according to whether they contain artifacts of stone, bronze, or iron. Thus, came what we know as the Stone Age, Bronze Age, and Iron Age that precede the ruins of the first historically known nations. The Scandinavian archaeologists knew that civilization had spread to their own land from the lands of the ancient Orient but had little difficulty fitting these stages well within the about six thousand years era of man as can be calculated from the Bible.

Then something happened to vastly expand the age of man. This was the shift of dating archaeological prehistory from an archaeological to a geological basis as I mentioned above which so expanded the antiquity of man just as I will explain in Chapter 30. From the beginning of geological science in the late eighteenth century, geologists ignored human remains, which were then supposed a matter for the antiquities of history rather than of geological science. When human remains were first considered by geologists, which took place in the latter half of the nineteenth century, geological dating had already been established independently of the remains of man which would have otherwise provided a far more accurate measure of the earth's most recent eras of geological history understood as being the Pleistocene (the Ice Age) and the Holocene, the latter being the present era of geological history dating from the end of the "Ice Age."

I mentioned in the *Archaeological Evidence* what clued me to the great problem in prehistoric chronology.[13] It was the *Glacial Kame* prehistoric burials in America's upper Midwest which are essentially identical to the Hopewell burials along the Ohio River (see Fig 12 next page).

Archaeologists recognize the same culture is responsible for these burials, but the former are found under kames, supposed to have been deposited by glaciers during the Ice Age. The Hopewell burials which are part of the larger Adena culture are currently dated to around the time of Christ but the lowest burials in these mounds are of the much earlier Archaic cultures of America. The upper levels of the Hopewell mounds which were used as burials by the historic American natives do date from the time of Christ though being of an entirely different material culture than the burials near the bottom of the mounds. Only the upper levels contain bow and arrows and sand-tempered pottery

---

[13] Philip Williams, *The Archaeological Evidence of Noah's Flood* (2011, CLSP), p 194

Why do Ice Age glaciers cover Archaic archaeological sites?

GEOLOGICAL EVIDENCE OF WORLDWIDE FLOOD HIDDEN IN PLAIN SIGHT!

Fig 12 - Antediluvian houses buried under "glacier" kames

which were used by the historical American natives. Because the historic American natives were accustomed to burying their dead in temple mounds, the top of these mounds would have been reasonable places to bury their dead if these mounds already existed, as had been the case since the time of the Flood. But why would ancient Americans living in Wisconsin have gone to the great trouble of digging beneath Ice Age glacier deposits to build elaborate houses for burying their dead? And why would ancient Americans living in Ohio have spent vast amounts of time and resources to build elaborately furnished mausoleums only to spend so much effort and time to cover them with dirt! That makes no sense.

The truth is, the mounds in both US states were created in the same way as the tels in the Ancient Near East, which is from eroded soil and gravel from Noah's Flood! The Paleolithic and Archaic cultures of America were contemporary and ended by the Flood in the third millennium before Christ. So were the Paleolithic, Epipaleolithic, Neolithic, Chalcolithic, and Early Bronze Age cultures all across the world also ended at the same time. Though imagined as belonging to successive periods, all these cultures were more or less contemporary. Rather than being from such vast expanses of time, all these sites and settlements were simply of different wealth due to being either rural or urban dwellings or else temporary hunting sites. All these sites belonged to the antediluvian world that was ended by the Flood. None may be older than about 6,000-10,000 years!

Part III

Discovering the Diluvium Flood and How it was Interpreted as a Long Ice Age

10

The Discovery of the Diluvium Flood

We should not be surprised to discover that Ice Age deposits cover prehistoric dwellings ended by the Flood because these Pleistocene deposits now supposed as the product of Ice Age glaciers were originally understood as evidence of a great deluge! As indicated by the name by which these Pleistocene deposits were first described, the *Diluvium* which we now attribute to a supposed Ice Age was originally understood as the deposits of what was perhaps a worldwide flood. Not all geologists agreed that the Flood that deposited the Diluvium was the same as Noah's Flood of the Bible or even that it covered every mountain. But most agreed with the world's leading naturalist Georges Cuvier (1769-1832) that some Flood as vaguely remembered in the accounts of many nations was responsible. *The American Journal of Science,* America's oldest and long its most prestigious scientific journal, was founded upon Diluvium geology.

Development of a Stratigraphic Understanding of the Earth

Discovery of the Diluvium required a stratigraphic (i.e. layered) understanding of the earth and a geological history of life. The first detailed understanding the earth's successive strata (c.f. Fig 13, next page) was developed by the founders of geological science such as Cuvier. This stratigraphic understanding of the earth however came about due to viewing the history of the earth in the light of the Bible, the only known history of the earth.

The principles of stratigraphy were put forth by the Danish naturalist Nicolaus Steno (1636-1686) who supposed that the earth was no older than the six thousand years that chronologists were calculating from the time of Adam. Noah's Flood was supposed as being the chief cause of whatever historical changes might be seen in the earth's strata. Against the teachings of the Aristotelians that petrified fossils were special forms of stones, Steno attributed petrified fossils to Noah's Flood. However, as I will explain below, it was the discovery of the true deposits of Noah's Flood near the surface of the earth that convinced geologists that petrified fossils as found in the earth's mountains were much older than man. By developing a stratigraphic understanding, geologists also discovered evidence of the earth's Creation, including the sudden appearance of life. Against the science taught by the Aristotelians and Deists, this was seen as a vindication of the Bible's account of Creation.

This earliest stratigraphical understanding of the earth's history was due to envisioning the continents rising above sea level due to a receding ocean, an understanding based on the ninth verse of Genesis.

And God said, "Let the water under the sky be gathered to one place, and let dry ground appear." And it was so. God called the dry ground "land," and the gathered waters he called "seas." And God saw that it was good. [Gen 1:9-10]

The cause of the receding oceans as taught by the German philosopher Gottfried Leibniz (1646-1716), was supposed due to the collapse of underground caverns as earlier proposed by Rene Descartes (1596-1650) and other thinkers from the Scientific Revolution. The oldest rocks were thus presumed to be those such as granite found near the tops of the mountains.

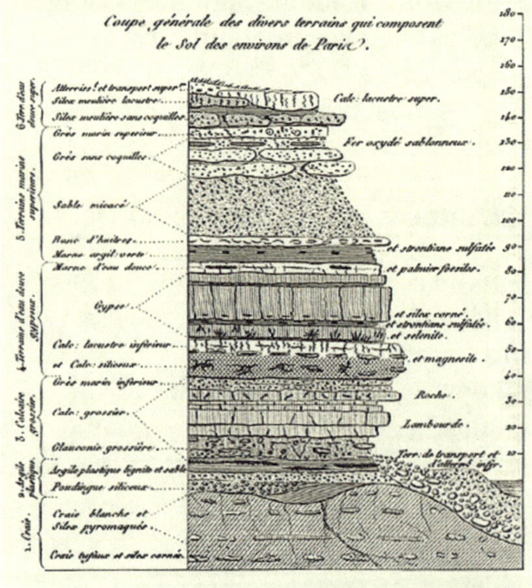

Fig 13 – Alexandra Brongniart's Illustration of geological strata as taught by Georges Cuvier

The Italian mining engineer Giovanni Arduino (1735-95) divided strata into four successive general orders which he named *Primary, Secondary, Tertiary,* and *Quaternary*. Arduino was building on the work of several German and Italian geologists who were studying the earth in the light of the Bible. These strata could be causally as well as historically ordered. The *Primary* or first rocks to appear from the ocean was supposed as being granite and thus found in the highest mountains. Due to lacking regular strata and having been formed from crystallization, *Primary* rocks were believed to date from the earth's Creation. Rocks containing strata reckoned to have been formed from the sand and sediment of eroded Primary rocks defined the *Secondary* strata. Fossils of mollusks, fish, and dinosaurs are found in the upper regions of the more recent Secondary strata. The *Tertiary* strata containing fossils of mammals and found along river valleys were originally believed to have been animals who perished in the Flood. Due to the discovery of the *Diluvium* in low lying regions, the Flood was understood as rather appearing in the topmost *Quaternary*, which are washed up deposits lacking regular strata.

The drawing of Fig 13 was published by the geologist Alexandre Brongniart (1777-1847) who studied the Parisian basin with Cuvier. In this drawing, the Diluvium lies piled next to the bottom of a cliff composed of the layers of rocks from the earth's earlier history. Though drawn as a valley cliff, the drawing is a composite of strata which Cuvier and Brongniart had discovered at various sites around Paris.

Though no evidence of man had been found in any of the strata, the fact that the Flood had been remembered by the nations caused Cuvier to believe that mankind was already living on earth at the time of whatever Flood deposited the Diluvium. Unlike Aristotelian and most Deist philosophers who presumed that mankind was as old as the earth, geologists had little difficulty incorporating a long history of the earth prior to the time of man. In contrast to the teachings of Aristotle, the Bible not only teaches the Creation of the earth but teaches the earth as having been created prior to man, the last of God's Creation.

The Diluvium deposits which lay near the surface of the earth had no well-ordered strata. In the words of a geological historian, it consists of

a chaotic assemblage of sediments ranging from muds, silts, sands, and gravels, which frequently had contorted structures, to extremely poorly sorted deposits we now refer to as till, with pebbles and boulders of rock embedded in a fine-grained matrix. A simple bedding in such deposits was frequently impossible to discern. [A. Hallam, *Great Geological Controversies,* p. 87]

The Diluvium is covered either by a layer of topsoil or else by the Alluvium strata from ordinary river floods that have formed after the Diluvium was deposited some few thousand years ago. The Diluvium also includes rich black soil called *loess* which extends across the earth's most fertile lands. Contrary to the current view of the loess having been deposited by "Ice Age" winds, it contains evidence of having been deposited by water. The Diluvium deposits also contain the bones of extinct animals which aside from their larger size and different geographical distribution are similar to living animals.

One can inspect the Diluvium by digging through the layer of topsoil that covers unimproved lands everywhere except at places like the eroded tops and sides of mountains. The Diluvium deposits, which are found near the surface of the earth, lie above the deeper and stratified Primary and Secondary rocks including those which contained the fossils of fish and dinosaurs. They also lay just above the Tertiary deposits which lie mostly along river valleys and coasts and contain the bones of early mammals. But the Diluvium lies just below the topsoil or else below the *Alluvial* strata created from the seasonal floods of present-day rivers, both of which contain the bones of current species. The Diluvium covers great valleys to various depths as might be expected from the deposits of a great Flood. But unlike either the river strata (the Alluvium) or the deep Primary, Secondary, and Tertiary strata that lie beneath the Diluvium, the Diluvium has no simple bedding pattern. Only a great Flood could explain such chaotic deposits which mark the sudden extinction of the earlier wildlife of Europe and America.

PROPOSED CAUSES OF THE DILUVIUM FLOOD

Though the source of the Diluvium seemed to have been some great Flood, geologists did not agree concerning the cause of the Flood that deposited the Diluvium. The great Swiss geologist Jean DeLuc attributed the Diluvium Flood to a sudden and catastrophic exchange (a "revolution") between earth and sea, a view that would be adopted by Cuvier. But the leading German naturalist Peter Simon Pallas and the leading French geologist D.G.D. Dolomieu believed the Diluvium was created when giant tidal waves either carried arctic birds south or swept tropical animals north. Among the animals which this Flood made extinct in Europe were elephants, hippopotamuses, rhinoceros, hyenas, and apes. At the same time, horses and camels which then roamed in America suddenly went extinct.

11

Geological Science from the Time of the Scientific Revolution

Understanding how geologists discovered the Diluvium Flood requires learning how philosophers and theologians had understood the science of the earth from the time of the Scientific Revolution. A theory of the formation of the earth apart from the rest of the universe did not make a lot of sense so long as people held to the Aristotelian teaching that the earth was the center of the universe. Thus, theories about the formation of the earth did not exist until after the Copernican Revolution when the earth was understood as a planet orbiting the sun and the sun as being a star in the greater cosmos.

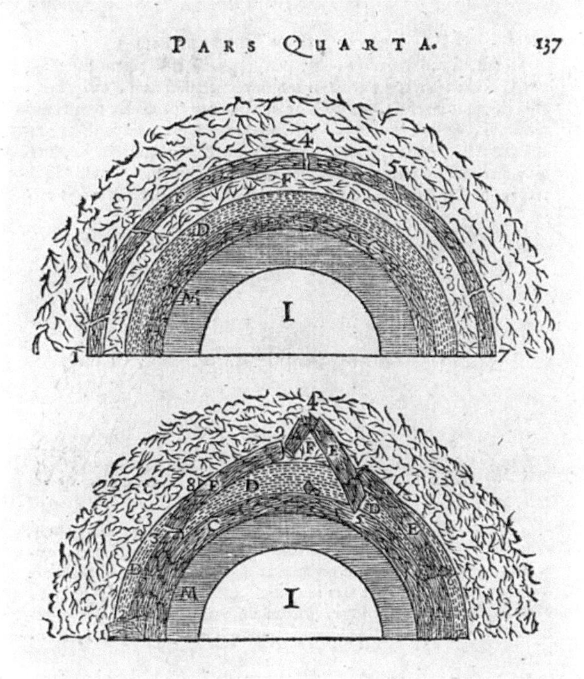

Fig 14 – Descartes Theory of the Earth

The first to put forth an explanation of the origin of the earth according to the new science of mechanics was Rene Descartes (1596-1650). Descartes explained the earth as having formed from a cooling star solely through natural processes (Fig 14). Likely to avoid censure as happened with Galileo, Descartes avoided any reference to the Bible, claiming this to be but an exercise in mathematics. Even so, Descartes contemporaries recognized that his "impious hypothesis" avoided the supernatural acts of Creation as taught in the Bible. Descartes claimed that miracles suggest that God's original design of the world was imperfect. Though his explanation of the formation of the earth excluded any mention of the Flood, Descartes diagram of the earth (Fig 14) included an internal layer of water and layers of internal caverns. His second diagram showed how the collapse of the internal caverns would lead to the forming of mountains.

Fig 15 – Gilbert Burnet's *Sacred Theory of the Earth*

Descartes theory of the earth inspired Bishop Gilbert Burnet (1643-1715) to write *The Sacred History of the Earth* showing how Noah's Flood happened as the result of the earth's ongoing processes. As illustrated on the cover of his "sacred" theory, Burnet claimed that the earth's ordinary geological processes followed biblical history, the earth having begun in (1) chaos before turning into (2) paradise. Having decayed due to sin, the water below the earth results in (3) Noah's Flood. The latter in turn brought about (4) the earth's present form including mountains and oceans. The present earth will however end with the (5) conflagration due to the Second Coming of Jesus, but (6) return to Paradise for the millennium reign of the saints. The millennium will then end with (7) fire from heaven which causes the earth to turn back into a star. Burnet's book included many detailed scientific observations supporting his grand theory. According to Burnet, the entire cycle of the earth's history would take place over a period of about 6,000 years. In Burnet's day that seemed a sufficient time for the history of the world.

Regardless his following the pattern of biblical history, Burnet's contemporaries recognized that by altogether avoiding any special or miraculous acts, his explanation of the earth's history through ordinary physical processes left no role for the acts of the God of the Bible. As was becoming fashionable with philosophers and theologians, Burnet had just rationalized the great biblical miracles including the Creation and Flood.

Though he admired Burnet's theory of the earth, Isaac Newton focused his attention on the creation of the heavens, as might be expected of one who used his new understanding of the movements of the stars to study ancient history and biblical prophecy. According to Newton, the chaos by which God began his Creation consisted of the matter of the solar system which the Spirit of the Lord distributed between the sun and the planets. The fixed stars were placed at a sufficient distance from one another to prevent their collapse when the Lord added gravity to these various collections of matter. Adding gravity caused the sun to begin to shine, forming the light on the first day of Creation. Clouds created by the division of the waters on the second day of Creation blocked the light of the sun until the fourth day of Creation when the earth was given its daily rotation. Newton's interpretation of Genesis continues to be used by many Creationists.

According to Newton, prior to the earth having been given a rotation equal to 365 days for each solar orbit, the days of Creation may have been of one-year duration due to the earth’s annual orbit of the sun. Limiting his theory to the Creation narrative, Newton gave no explanation of Noah's Flood. Rather, he saw God as a Ruler who supervises his Creation by adjusting the orbits of the planets which were endangered by their mutual gravitational attraction. Thus, did Newton propose what became known as the 'God of the gaps,' the practice of using supernatural acts of God for that which could not be explained by ordinary natural processes. These acts of the 'God of the gaps,' are not specifically mentioned in the Bible. But Newton believed that the Jews and Catholics had corrupted the Scriptures. Newton explained the narrative of Creation given in the Bible as that of a man standing on the surface of the earth, an accommodation for 'the vulgar masses who are incapable of a nice philosophical understanding.'

William Whiston (1667-1752), Isaac Newton's chosen successor for chair of mathematics at Cambridge University, used comets to explain both the creation of the earth and Noah's Flood. According to Whiston, the earth was formed from a giant comet which had gotten too close to the sun. He explained the Flood as coming about due to God sending another comet to approach the earth. Following Newton's suggestion, Whiston taught that 24-hour days came about due to the Lord causing the earth to spin on its axis on fourth day of creation. Thus, Whiston taught that the first four Creation days were each one-year in length, day and night then coming from the earth circling the sun.

Cambridge University's John Woodward (1665-1728) insisted that more attention ought to be given to the empirical data of earth such as in the distribution of fossils. Woodward's collection of fossils became one of the first museum collections by which the earth's history would be studied. Following Isaac Newton in seeing gravity as something that God added to matter, Woodward suggested that the Flood was caused when God temporarily suspended the law of gravity. The fact that the lowest lying fossils are shellfish is due to their higher specific gravity as God reenacted the law of gravity after the Flood. However, the attention that Woodward gave to the strata of the fossils would lead to geologists recognizing that their disposition could not have been from a single Flood, an opinion already recognized by the Puritan naturalist John Ray (1627-1705).

Though he had originally proposed a steady-state model of the earth, France's leading naturalist Comte de Buffon (1707-1788) was inspired by Whiston's extended duration of the days of Creation to develop a model of the earth as principally due to the cooling of the earth from a hot star over seven epochs totaling 75,000 years. Buffon based his estimate on the time required for the cooling of liquid iron, which he determined by his own experiments. Buffon considered but did not embrace a theory of the evolution of species.

Though recognized as the world's leading naturalist, Buffon's shift from a steady state earth to one that reflected the biblical days of Creation disappointed the Deist Voltaire, who preferred Whiston's theory of the earth due to being based on Isaac Newton's mathematical principals of natural philosophy. Voltaire liked the fact that Isaac Newton distinguished his 'nice philosophical explanation' of Creation from the account of Creation in the Bible, which Newton taught as having been 'intended for the vulgar masses.' Following Newton, Whiston believed that Moses had learned the stories of Creation from the Egyptians which the 'rude and illiterate Jews' had corrupted. He would not rule out the possibility that the Egyptians had been spared the Flood.

This was now the time of the Enlightenment when the truths of *reason* would be extolled above the narratives of the Scriptures, thus giving rise to Deism. Though Newton and Whiston's preoccupation with biblical prophecy kept them tied to the Bible, and though both men believed that God used miracles to create the world and send the Flood, Newton's mathematical explanation of the world served to inspire Deist views in which miracles would be excluded from science.

These are hardly the only theories of the earth's creation appearing in the eighteenth century, but one can understand why by the end of the eighteenth century, many began seeing an empirical investigation of the earth's minerals, fauna, fauna, and fossils as being more valuable than such grand systems of the earth's creation. However, another reaction was to altogether reject theories of the earth's creation and formation as would be the strategy of the Deists.

12

Deism's Important but Neglected Role in the Rise of Science

Essential for understanding the development of geological science and opposition to the Diluvium Flood is the role of Deism. Deism became the worldview of secular leaders and thinkers from the eighteenth-century Enlightenment until the rise of atheism over the course of the nineteenth century. Many have heard of Deism and know that some of American founding fathers were Deists but do not well understand the theology of Deism nor its important role in the foundations of Science.[14] Because Deists recognized a Creator and because most of them were not hostile to Christianity, many Deists have been supposed as having been Christians. Indeed, most Deists were nominal Christians and members of Christian denominations. This is particularly the case of America's Founding Fathers. But Deists reject miracles and spirits and therefore reject a literal reading of the Bible.

Creationists may be surprised to learn that Deists such as Voltaire who so despised Christianity and the Bible, loved Intelligent Design because it was a revelation of the Creator apart from the Bible. But secularists may be surprised to learn of Deism's essential role in establishing the intellectual foundations of modern science which happens to be the Intelligent Design of the Deist Creator! This is due to an atheistic Darwinism having replaced Deism as the chief opponent to the teachings of Christianity and the Bible.

The two essential elements of Deism are (1) the rejection of miracles, divine revelation, and unseen spirits such as angels and demons which are the supernatural elements of the Bible, and (2) the rejection of a literal reading of the Bible especially the Bible's testimony

[14] I capitalize *Science* whenever it is an idea of devotion rather than the practice of seeking knowledge concerning the physical world.

of miracles and divine revelation. However, even Deists such as Voltaire who were most hostile to the Bible and Christianity regarded the Bible as useful if not essential for teaching morality to the unlearned masses. Many understand that Deists believe in a distant Creator God who does not involve himself in his Creation. They may also be aware that Deists believe in an afterlife where God reward's people for their good deeds and punishes bad deeds. In truth, Deism is not so different from liberal Christianity which has been recently described as *moralistic, therapeutic* (due to sickness replacing sin), *deism!* Deism was also the reason that moral science replaced biblical theology in many universities. Nor is Deism so different from those who regard themselves as orthodox Christians but employ composition criticism to suppose a rationalist (non-miraculous) explanation of the Bible.

The rise of Deism was due to the success of the Scientific Revolution in explaining the natural world including things that had previously been attributed to miracles or unseen spiritual powers such as angels. Though an avid student of biblical prophecy, Isaac Newton is much responsible for popularity of Deism by explaining physical motions in both the heavens and earth in terms of the mathematical principles of natural philosophy.

Even more important is the fact that Newton was almost singularly responsible for what philosophers call the *disenchantment* of the world, the elimination unseen spirits and a non-physical soul which survives the death of the body. **Newton moved the world from the premodern to the modern worldview** by interpreting the (1) the angels of the Bible as human messengers, (2) devils as bad human actors, (3) and witches as dispensers of drugs. This led to allegorical interpretations of the supernatural accounts in the Bible. Even more important was (4) Newton's teaching of *mortalism,* a reference to the souls of the departed as remaining in the grave until a future resurrection. This challenged the Christian belief that dying Christians go immediately to Heaven.

Likely, Newton was influenced by the Aristotelian writings of the Rabbi Moses Maimonides whose writings were among Newton's private papers. Thus, it would not be inaccurate to see the worldview of both Newton and the Deists as more Jewish than Christian if for no other reason than their rejection of the deity of Jesus. Newton did however believe that Jesus's death atoned for man's sin, that Jesus resurrected from the dead, and that he would return to the earth shortly after the second millennium to rule from a rebuilt temple in Jerusalem.

Newton is thus much responsible for the famous disappearance of the soul apart from a physical body which led to a materialist or physicalist view of man himself which conflicts with the spiritual understanding of man taught in the Bible. One of the first scientific psychologies was soon developed by David Hartley (1705-1757). Newton thus emptied Heaven of everything but God and taught history as the work of human beings aside from a single providential deity who was not Jesus Christ.

Though Newton spent much of his effort calculating the Second Coming of Jesus, he supposed the deity of Jesus a Catholic invention. These teachings, only later discovered in his private papers, were hardly known to most of his contemporaries but were spread through freemasonry after the triumph of the ancients over the moderns in this secret society. Freemasonry thus created the Enlightenment among the social elites of Europe and America. Most important, the disenchantment of the world introduced by Newton marks the shift from the premodern to the modern age!

Isaac Newton is a complicated figure and understood in different ways even by his contemporaries, but the popular way that Voltaire introduce him to France and Germany as a great man who discovered nature's basic laws was much responsible for the popularity of Deism. The place of Isaac Newton in the Enlightenment is not unlike how atheists would come to celebrate Darwin.

While there were Deists such as Voltaire and Thomas Paine who despised Christianity, most Deists continued to regard themselves as Christians and were members of churches as in the case of many of America's Founding Fathers. Deists referred to their Deity as *Nature's God,* as the *Supreme Creator,* or as *Providence,* but never as Jesus Christ! Unlike the Aristotelians, Deists taught that the world was created. But contrary to the Bible and historical geology, Deists assumed that the earth was fully formed in its current state from the time of its creation. They also believed that the earth was providentially created for the benefit of mankind. Thus, they had a genuine fondness for Nature's God, for whom they preferred the name *Providence.*

Similarities between Deism and Young Earth Creation Science

In supposing an instantaneous Creation, Deist science is not so different from that of the Young Earth Creationists who teach that God fully created the earth and everything in the earth in only 144 hours, leaving little if any evidence of an historical record apart from the fully formed Creation. As the case of today's Young Earth Creationists, Deists supposed that God created the world in a fully mature and functioning state. Nor did Deists have a problem with a young earth. Voltaire accepted Archbishop James Ussher's dating of Creation to 4004BC. Of course, even if they teach a science which has much in common with that of the Deists, Young Earth Creationists do aim to literally believe the Bible.

Moreover, Deists could equally believe in an earth that is indefinitely old, essentially the same view as the Aristotelians who taught that the earth and mankind have always existed. Some Deists who opposed the Diluvium Flood tended to make an exception for the ephemeral history of man which they separated from the science of the earth. Deists separated the truth pertaining to the earth which they believed could be mathematically understood from the Bible which they assigned to the genre of history and literature. This is similar to the teachings of Immanuel Kant who so influenced liberal Christianity. Because truth is also a moral issue, liberal teachings which deny the historicity of the Bible's miracles are not only contrary to a literal reading of the Bible but are an inconsistent theology and moral philosophy. If the Bible does not give us the truth about history, it belongs to the genre of myth which the Bible's own teachings so deplore and condemn.

13

Opposition to the Diluvium Flood

The geologists who discovered the Diluvium Flood as well as a progressive creation of the earth were opposed by disciples of the Deist James Hutton (1726-1797), who popular geological histories herald as the first *scientific* geologist. This was certainly Hutton's aim. As Isaac Newton had discovered laws governing the system of the heavens, Hutton claimed to have developed a system that could explain the earth in a quasi-mathematical way. A friend of James Watt who invented the steam engine, Hutton taught that the earth itself functioned as an eternal machine. Denying that the earth revealed 'any evidence of a beginning or prospects of an end,' Hutton rejected both evidence of the earth's Creation as well as a worldwide Flood, things that would be an exception to laws governing the earth's ongoing geological processes.

Geologists would eventually reject Hutton's uniformitarian geology but, as I will explain in Chapter 21, his celebrated role in the history of geology, stems from the resurrection of his views by another Deist, Charles Lyell, the celebrated mentor of Charles Darwin. Hutton did have an important role in the history of geological science, though mostly as a foil for those who developed historical geology which, regardless its biblical roots, would be essential for Darwin's theory of evolution. As the case of his contemporary Scotsman, David Hume, Hutton wrote against the miracles of the Bible as violations of the laws of Nature. Most famously, Hutton taught that unlimited amounts of time were available to explain the earth's processes. This was not however to propose an old earth, but rather to deny that the earth revealed any evidence of a beginning.

Hutton's System of the Earth

Hutton supposed that the earth's processes are cyclical, as the case of the earth's current hydrological (rainfall) cycle to which Hutton made a significant contribution. Hutton explained how water from the earth's springs, which empty into streams and rivers, flow into the oceans. The springs however are replenished by rainfall that comes from the clouds formed from the evaporation of the waters of the oceans.

More famously, Hutton also proposed that erosion due to rainfall entirely erodes the earth's continents, carrying topsoil into streams and rivers which empty into the oceans. Thus every mountain and continent

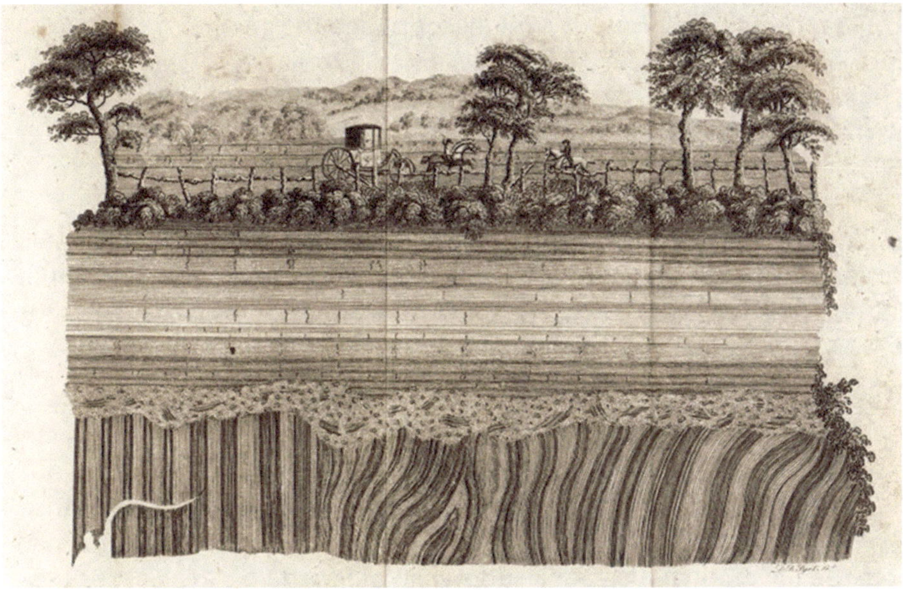

Fig 16 – James Hutton's Illustration of Unconformity in Scotland

will eventually erode to sea level. As the weight of the eroded sand builds up in the ocean basins, the pressure of this weight forms rocks in horizontal strata as depicted in his famous illustration (Fig 16). As the number of horizontal strata increase and sink due to the weight of ever more sand flowing into the oceans, the horizontal strata will eventually melt to become basalt which is extruded to the surface by heat deep within the earth. Just how this happens, Hutton could not explain, but these extruded rocks form vertical columns that push up the horizontal strata that lie closer to the earth's surface. Hutton denied that there existed "Primary" rocks belonging to the earth's creation. Rather, he taught that the Primary rocks are constantly formed by the melting of the horizontal strata.

Due to Darwin's theory of progressive evolution and more recently having come to believe in catastrophes of sudden extinctions, geologists no longer hold to Hutton's uniformitarian view of the earth. Unfortunately, his teaching of the volcanic origins of the earth continues to inform current geological science causing much confusion in geological science, especially as would be used by Charles Lyell to try to explain away the Diluvium Flood.

DISCIPLES OF ABRAHAM WERNER VERSUS DISCIPLES OF JAMES HUTTON

The disciples of Hutton famously opposed the geognosy of the German mining engineer Abraham Werner (1749-1817). Geognosy refers to the positions of rocks, not their history. This may have been Werner's way of avoiding historical geology because Werner was no less a Deist than Hutton. But Werner's system of rocks being formed due to a diminishing ocean provided a historical understanding to the earth similar to the Bible's account of the earth's formation. British disciples of Werner included the Irish chemist Richard Kirwan (1733-1812) and Robert Jameson (1744-1854), Regius Professor of Natural History at the University of Edinburg and founder of the Wernerian Natural History Society.

Due to teaching the origin of rocks in the oceans, Werner's disciples would be called *Neptunists,* after the Greek god of the sea. Hutton's disciples were accordingly called *Plutonists* after the Greek god of the fiery underworld. Whereas Werner explained the continents as formed from the lowering of the seas, Hutton explained the earth's process as driven by a fire deep within the earth, the same as responsible for volcanoes and the columnar extrusions at the bottom of Hutton's drawing (Fig 16). While Hutton recognize basalt as having been formed from volcanoes, Werner incorrectly taught that basalt was formed by a chemical process in the deepest part of the oceans. Though otherwise holding to his teachings, increasing numbers of Werner's disciples acknowledged that basalt was formed by volcanoes, causing them to be called *Vulcanists* after the Greek god known for forging of iron. Hutton's acolytes make this distinction in recounting the history of geology so as to discredit Werner's essential role in the rise of geological science by developing the first stratigraphic understanding of the earth's rocks. The truth is, granite is indeed crystalized in water, though not in the oceans but in the far greater pressures that exist deep within the earth.

BASALT VERSUS GRANITE AS THE EARTH'S PRIMARY BASE ROCK

To explain the earth's processes as a great cycle, Hutton supposed that granite was somehow formed from the melting of volcanic basalt deep within the earth, then extruded as veins of granite inside basalt, which he supposed as being the earth's fundamental bedrock. Ironically, the disciples of Hutton would use the earth's scattered and unpredictable volcanoes to explain away the Diluvium Flood, which Hutton's disciples disliked precisely because it could not be explained as part of the uniform operations of the earth. However, they needed volcanoes to explain the rise of both mountains and continents as would later be the case of Charles Lyell who would resurrect Hutton's uniformitarian geology.

Interesting that Young Earth Creationists use volcanoes to defend their young earth view, as in the case of Mount Saint Helena in Washington State. In their case, it is because volcanoes can rapidly create strata.

GRANITE AND WATER FOUND DEEP INSIDE THE EARTH

The base rock of the earth's ocean basins are indeed made of basalt. However, the Kola Superdeep Borehole, drilled more than 7 miles deep in northern Russia failed to find even scattered veins of basalt in the earth's deepest rocks, but only the lighter granite. Deeper than 3 miles, they found increasing amounts of water percolating from below until blocked by layers of impermeable rock. In Chapter 57, I will explain this water was responsible for the hydrological cycle prior to the Flood that was different from how the earth is watered today as also the Flood itself.

14

Jean DeLuc and the Turn to Geological Empiricism

Rather than reject theories of the earth's origins as did James Hutton, another response to so many competing speculative theories was to reject all speculative theories including the one of James Hutton for a modest empirical study of the features of the earth. The leader of the move towards geological empiricism, thus the development of historical geology as also the Diluvium Flood, was Jean DeLuc (1727-1817) who due to the latter role has been much neglected in popular histories of geological science. DeLuc insisted that *geology,* a word which he popularized, be limited in scope and be based on an empirical study of the earth.

The DeLuc family of Swiss watchmakers were known for defending the Bible against Deists such as Voltaire who lived in a French town near DeLuc's home in Switzerland. Against the persuasions of Voltaire, DeLuc counseled Rousseau to return to his Reformed faith. DeLuc was also a scientific instrument maker who invented the portable mercury-based barometer for measuring atmospheric pressure. He is also credited with improving the gas laws regarding water vapor. After a business failure, DeLuc moved from Switzerland to London to serve as Reader for Queen Charlotte, wife of King George III.

DeLuc may have looked forward to the move because, against the *a priori* theoretical tradition of the Continental philosophers which had begun with Descartes, DeLuc championed empirical science, long associated with British philosophy. Throughout the Middle Ages, empiricism was associated with British champions of the Bible against the speculation of the Aristotelians. These British theologians did not want to see the acts of the God of the Bible limited by the God of the philosophers. Historians of science refer to this as *voluntarism* against a *philosophical* understanding of Creation as taught by Thomas Aquinas and disciples of Aristotle. The voluntarist view freed the God of the Bible from philosophical determined principles and was key to the development of the science of mechanics from the time of William of Ockham (1287-1327) to Robert Boyle (1627-1781) and Isaac Newton (1643-1727).[15] This radical view of the sovereignty of God also influenced Calvinist theology, though the reverse was also true.

Against the *a priori* reasoning of Continental philosophers such as Descartes, Leibniz, and Huygens, the British scientists Isaac Newton and John Locke (1632-1704) championed empiricism. Perhaps because Voltaire had also famously championed Newton and Locke, DeLuc championed Francis Bacon's (1561-1626) experimental science and inductive methodology.

Jean DeLuc's profound influence on geology is due to five important things: (1) his insistence that geology be based on *actual causes,* by which he meant the earth's presently-observable processes, (2) his recognizing the recency of the earth's rivers; (3) his recognizing a world before Adam, thus becoming the father of what is now called *Old Earth Creationism;* (4) his connecting geology with human history by pointing to the connection between archaeological ruins and geological processes; and (5) his teaching of the Flood as being caused by the revolution of land and sea.

Contrary to James Hutton, DeLuc believed that the empirical study of actual causes revealed that the earth's present processes to have been different in the past. These became key elements in the development of the Diluvium Flood though DeLuc's teaching of the Flood as having been caused by a revolution between land and sea would contribute to the eventual rejection of the Diluvium Flood. Moreover, DeLuc's belief that certain archaeological ruins date from shortly after the Flood may have been responsible for geologists failing to realize that the Diluvium Flood was also responsible for most prehistoric ruins.

---

[15] .c.f. Eugene M. Laaren, *Religious Origins of Modern Science* (1977, Eerdmans), Chapter Two: The Development of a Voluntarist Theology of Creation

15

The Two-Book View of Divine Revelation and the Baconian Compromise with Scripture

So radical was DeLuc insistence of empiricism that even as an orthodox defender of the Bible he also rejected the *a priori* use of Scripture in the practice of science. To defend his rejection of Scripture for interpreting the earth, DeLuc appealed to Francis Bacon's 'Two Book' view of God's Revelation. Bacon's two books are the Bible as 'the book of God's *words'* and the Creation as 'the book of God's *works*.' While Bacon taught an experimental investigation of God's Creation as a work of piety, he rejected using the light of Scriptures to understand the Creation.

Requiring the reinterpretation of the Scriptures to accord with the latest science has been called the *Baconian Compromise with Scripture*. The Two Book view frees science from the light of Scripture while it uses science to interpret Scripture. With regard to Creation, science becomes Scripture's master! Thus, it is hardly a high view of the Scriptures.

Such a compromise with Scripture was hardly true of the great thinkers of the Scientific Revolution such as Isaac Newton and Gottfried Leibniz who explicitly studied the Creation in the light of the Bible. Even so, these leaders of the Scientific Revolution were also reading the Scriptures themselves in the light of the science deriving from the Copernican Revolution as do even Creationists.

However, if the Scriptures are already being read in the light of some earlier science, as was the case of the Aristotelian earth-centered worldview and today's Young Earth Creationism, our reading of the

Scriptures may need to be adjusted by a fresh examination of the latest empirical discoveries in the light of the Scriptures.

THE FATHER OF DAY-AGE CREATION

Likely, DeLuc rejected the a priori use of Scripture due to expecting resistance to his new view of the earth as being far older than man. DeLuc was the first Old Earth Creationist and subsequently became the first Day-Age Creationist. DeLuc supposed that more information about the earth may be needed prior to interpreting the words of the Bible as pertaining to Creation and the Flood. Thus, DeLuc attacked the Irish chemist Richard Kirwan for his '*a priori* use of Scripture' for explaining the earth. This was perhaps due to Kirwan maintaining the view prevailing since the time of Bishop James Ussher that the earth was but a few days older than man.

James Hutton submitted his theory to DeLuc for review. Though he criticized Hutton's system, DeLuc was more respectful than in the case of Kirwan. One reason was because DeLuc had insisted that geology be based on *actual causes,* by which DeLuc meant geological processes that could be presently observed.

Hutton's system was supposed as being based on actual causes. However there were things in Hutton's system that can never be observed, for example his teaching that deep inside the earth, Primary Rocks are being constantly formed from melting of Secondary Rocks. Rather than being an induction from the observed changes in the earth, Hutton's system was imposed on whatever might be observed. The same would be true of the method of Charles Lyell who would champion Hutton's system. Giving empirical evidence that the present continents could not be so old as required by Hutton's system of the earth, DeLuc rejected Hutton's theory of the earth.

The problem with the Two Book view of Creation, which is embraced by some Old Earth Creationists today, is that there are no uninterpreted facts. Thus, we will always interpret God's works whether according to the narrative of Scripture or some scientific theory. Why should one *a priori* reject the light of the Scriptures which do have something to say about history and the Creation? If we reject the light of Scripture, we limit ourselves to some or another fallible human theory! DeLuc may not have been aware of how much his own theories depended on the light of the Bible. He did however believe that his empirical study of the earth independently of the Bible provided evidence both for the earth's

Creation and the Flood. That DeLuc discovered a world existing before Adam was due to examining the empirical facts in the light of the Bible.

Many suppose that the light of the Bible has increasingly disappeared from science for an entirely secular science. The opposite is the case. To the contrary, the light of the Bible has only slowly been used for understanding science. Thus, science has been increasingly influenced by the Bible. For example, the rejection of the eternalist view of Aristotelian science and the Deist scientific views being discussed here for the current historical sciences is due to the light of the Bible. So was the rejection of astrology, sorcery, and magic, which where were part of the older Aristotelian science. Science progresses as it becomes ever more influenced by the Bible.

For example, that the historical sciences even exist is due to studying the world in the light of the Bible, the root of all historical worldviews. As I will explain in Chapter 26, not only is our Christian faith, but also the Western world's linear sense of history is grounded in the historical miracles of the Bible as those of Creation, Noah's Flood, and Jesus's resurrection of the dead. History is not based on the passage of time but, as the case of its sudden creation and the appearance of life on earth, on contingent world-changing events as those put forth in the Bible.

DeLuc was a true empiricist. Most naturalists of his day limited their study to books, libraries, and museums. During a time when horseback, horse-pulled coaches, and wind-powered sailing ships were the chief means of travel, De Luc visited the mountains of his native Switzerland and the coasts of northern Germany, home of his patroness Queen Charlotte, wife of King George III. It was due to his firsthand observations that DeLuc led the way to the empirical geology that established the historical sciences.

16

River Deltas, Eroded Volcanic Cones, and Wide U-Shaped Valleys as Evidence of a Recent Flood

Challenging Hutton's system of proposing an indefinitely long era for the earth's processes, DeLuc pointed to the river deltas in the lakes of his native Switzerland. Hutton's system suggests these lakes should have become filled with silt long ago. The finite extent of river deltas where rivers empty into lakes and oceans proves that rivers cannot be so old as Hutton claimed. Thus, DeLuc insisted that the processes forming the earth's continent cannot have extended very far into the past.

De Luc was not the only geologist who recognized that wandering rivers with V-shaped banks cannot have excavated the wide U-shaped valleys through which these rivers flow. Some valleys lacked rivers and streams. Only such erosion and deposition as could have come from a great Flood could explain the wide U-shaped valleys. Thus, most geologists rejected Hutton's theory until resurrected by Charles Lyell.

As did the disciples of Werner, DeLuc recognized that the present continents once lay under the oceans. Evidence of this are marine shells found on land. What was new with DeLuc is his teaching that oceans also cover what was once land. This was indeed the case of the North Sea which once connected Great Britain to the continent of Europe. Though the sinking of this land below the ocean was due to the waters of the Flood, this was not due to a revolution between land and sea as DeLuc supposed.

STONE AGE DWELLINGS AND THE FLOOD

As evidence that land had only recently risen above the ocean, DeLuc pointed out that the peat that covered Stone Age mounds (see Fig 4 in Chapter 3) was of the same depth as the peat covering the coastal heaths. According to DeLuc, this showed that the exchange between land and sea occurred within the era of human history and not even at the beginning of human history but since the time of the Flood, which according to DeLuc was caused by a revolution between the continents and the ocean.

Because he believed that prior to the Flood, the present land lay beneath the ocean, De Luc could not envisioned this stone-age dwelling as having been built prior to the Flood. As I explained in Chapter 3, those who built and lived in these stone dwellings were in fact drowned by the Flood that deposited the peat on the heath as well as on these Stone Age dwellings. Contrary to what De Luc taught, these Stone Age dwellings were built before, not after the Flood!

REVERSALS BETWEEN LAND AND SEA

Here we see the great problem with DeLuc's explanation of the Flood as having been caused by a revolution between land and ocean. DeLuc however lived before archaeology had become a science. When it did become a science, these dwellings continued to be seen as having been built after the Ice Age ended. Even so, DeLuc showed how to connect archaeological remains with geological deposits, thus providing a way for connecting geology to human history.

The source of Flood waters sufficient to cover the world's highest mountains has long been a mystery. It is thus understandable that DeLuc and those who followed him would suppose the waters of the oceans as having been necessary for a world-wide Flood. However, the lack of marine shells found in the Diluvium and the shells of fresh-water mollusks mixed with marine fossils would eventually be used to refute the Diluvium Flood. But neither is a sudden exchange between land and the oceans how the Bible explains Noah's Flood. In Chapter 59, I will explain how the waters of the Flood did not come from the oceans, but exactly as explained in the Bible.

MODERN AND ANCIENT VOLCANOES

Supposed as more evidence of a recent Flood were the eroded cones of volcanoes which contrasted with the intact cones of more recent extinct volcanoes as those in France's Auvergne mountains. Many geologists explained eroded cones as distinguishing volcanoes which date from before the Flood from volcanoes dating after the Flood. However, all the volcanoes in the Auvergne Mountains had become extinct before any existing human records.

17

Erratic Stones as Evidence of a Recent Flood

Some of the most powerful evidence for a worldwide Flood, still the case to this day, does not come from the depths of the earth as Young Earth Creationists teach, but are stones lying on the earth's surface that are found tens or even hundreds of miles from their place of geological origin. How were these stones transported so far from the place of their geological origins? Such stones had been recognized at least from 1719 when Emanuel Swedenborg served as the Director of the Swedish Board of Mines in Stockholm, Sweden. The explanation of these stones was universally supposed as diluvian even after it became obvious that the deeper parts of the earth dated far earlier than man.

Another who recognize the displacement of these stones was the influential nobleman Benedict de Saussure (1740-1799), one of the first to climb Mount Blanc, Western Europe's highest mountain. Saussure was hoping to confirm Pallas teaching that the outcrops of rocks on the world's highest mountains were composed of granite. But he also recognized that stones lying in the Alpine valleys were of the specific granite that he had observed

Fig 17 – Erratic in the Burren, County Clare, Ireland

at the top of Mount Blanc. Though made of the same granite, these stones were many miles from the mountain, in some cases with an intervening valley. Not a few of these are precariously perched on smaller rocks (c.f. Fig 17). There being no local source to explain their presence, geologists referred to these displaced stones as *erratics*.

Erratics are of various sizes, ranging from small pebbles to some as large as houses. Some sit on land higher than the sources from which they originate. One might suppose that the stones were carried downstream by rivers, but their distribution did not just follow river valleys. Most of the large erratics had rough or ragged sides as those in Fig 17. But some had smooth surfaces as if they had long been exposed to running water prior to being displaced.

DeLuc suggested that the stones were shot across the land due to explosion coming from the collapse of underground caverns. But Saussure pointed out that the movement of such heavy stones would require the weight of rapidly moving water. Thus, Saussure attributed their placement to an extraordinary debacle as had recently occurred due to the failure of a natural dam high up the Swiss mountains. Dolomieu, on the other hand, guessed that the *erratics* had somehow tumbled from the mountain. He explained the intervening valleys as coming from later erosion. However, in a few cases, they were found on hills at a higher elevation than their source.

Fig 18 – Erratic rocks on coast of Estonia that crossed Baltic Sea from Scandinavia

Hutton and some of his disciples suggested that these stones were carried to their location by ancient glaciers that once extended from the Alps. His explanation obtained little attention and support inasmuch as slow-moving glaciers would be far less capable of moving these great rocks than would fast running water. The German geologist Christian Leopold Von Buch (1774-1853) noted that their placement was such that the *erratics* were carried through the air in the manner of cannon balls. His calculations required what he recognized as incredible velocities far greater than anything ever seen. Thus, the best explanation of the placement of these *erratics* was to thus to attribute them to a Flood far more extensive and powerful than anything ever seen in recent history.

Parallel groves found on many large erratics were explained as scratches made by smaller stones propelled across them in the same rapid currents as moved the erratics. Hutton's glacier explanation had even less support when it was discovered that not all *erratics* are found near mountains. The German Erhard Wrede (1766-1826) observed them on the northern coast of Prussia (Fig 18, next page). The only source for these stones were on the coast of Scandinavia across the Baltic Sea. But what could have transported these large stones across the Baltic Sea? Lamarck had suggested that there had been a tilt in the earth's axis which had lowered the sea level in the recent past. Wrede believed that the Baltic Sea was frozen before the oceans were suddenly lowered. He suggested that when that happened, floating icebergs carried the erratic stones across the Baltic Sea.

After Hutton's death, his theory was defended by his friends, the mathematician John Playfair (1748-1819) and the geologist Sir James Hall (1761-1832). Playfair wrote a popular summary of Hutton's theory of the earth avoiding Hutton's Deistic language to make it more acceptable to Christians. Hutton's disciples wanted to ignore the problem of *erratics,* but Hall had observed them on a visit to Switzerland. He pointed to grooved *erratics* that could be found just west of Edinburg's Castle Rock upon whose bedrock could be found grooves pointing in the direction of the *erratics*. Thus, Hall sought a way to incorporate the *erratics* so as to see them as being part of the earth's ongoing processes.

Following Wrede, Hall suggested that *erratics* had gotten frozen into the ice at a time when the North Sea was completely frozen. According to Hall, earthquakes which Hutton taught as giving rise to continents also occurred in the middle of the oceans, causing tidal 'waves of translation' far greater than any ever seen. With the lift of the floating icebergs, these tidal waves swept great erratic stones across Edinburg forming grooves on the bedrock before depositing the *erratics* just west of Edinburg.

Hall's highly speculative *ad hoc* theory violated the principle of actual causes due to being based on a phenomena that no living person had ever seen. Even so, Hall made tsunami Floods caused by what he called 'waves of translation' part of a supposed regular operation of the earth. However, calling these 'waves of translation' *diluvian waves* just gave all the more attention to the problem of the erratic stones which confronted the disciples of James Hutton.

18

Mass Extinctions as Evidence of a Recent Flood

In the years following the French Revolution, the most influential advocate of a recent geological Flood was Baron Georges Cuvier who I mentioned in Chapter 10. Cuvier had succeeded the famous eighteenth-century naturalist Comte de Buffon on the staff of France's Museum of Natural History, the world's wealthiest and most prestigious museum. Such government funding of geology resulted in the French taking the lead from the British naturalists in the development of this new science.

Cuvier founded the science of paleontology by teaching biologists how to reconstruct the appearance and behavior of extinct animals from a single surviving bone. So famous was Cuvier's reputation that fossil bones were sent for him to examine from places as far away as Russia and America. Against the evolutionary teachings of his equally famous colleague and rival Jean-Baptiste Lamarck (1744-1829), Cuvier famously established the reality of extinctions. A recent worldwide Flood was essential to Cuvier's argument for extinctions.

Naturalists were long familiar with the fossils of what we now understand as extinct species. But in Cuvier's day, it was not clear whether such species still existed some distant place on what was then a mostly unexplored earth. Another possibility was that the species of unfamiliar animals had just evolved into the familiar kinds of living animals as then being taught by his rival Jean-Baptiste Lamarck. While Lamarck focused his study on invertebrates, Cuvier studied the extinct vertebrate found in the more surficial deposits of the nearby Paris basin. To do this, he worked closely with the minerologist Alexander Brongniart who is responsible for the stratigraphic drawing in Chapter 10 above.

The bones of large mammals such as elephants, rhinoceroses, and bears were key to the question of extinctions because should there be any living mammoths, giant cave bears, or saber-toothed tigers, these large animals were the least likely to have been hidden from adventurers and explorers of the earth's vast wildernesses.

Part of Cuvier's argument against Lamarck's theory of evolution rested on the embalmed animals that Napoleon's troops brought back from Egypt. Cuvier pointed to the fact that these animals were remarkably similar to existing species. If these animals had not changed in several thousand years, it was unlikely that species could have evolved in the short period that had passed since the extinction of the large vertebrate animals that we now call megafauna. Such megafauna, which included numerous tropical species once roaming Siberia and northern Europe seem to have suddenly disappeared with no fossil record of them having co-existed with, much less having evolved into Europe's current species.

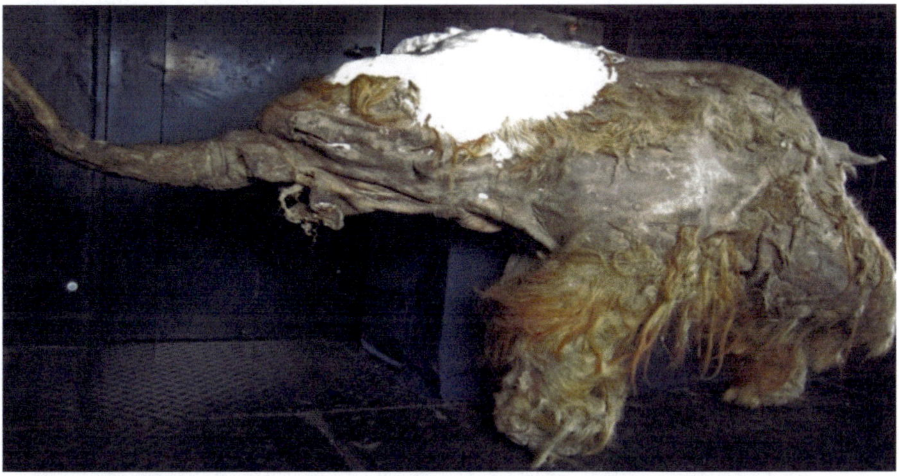

Fig 19 - Frozen carcass of the extinct woolly mammoth

Cuvier was also aware of the frozen carcasses of wooly mammoths and rhinoceroses that the Prussian savant Peter Simon Pallas (1741-1811) had reported as having been found in the tundra of Siberia (Fig 19). The meat of these carcasses was so fresh as to have been recently eaten by wolves. Following Pallas, Cuvier explained the sudden disappearance of what we now call megafauna as due to a recent worldwide Flood. As Cuvier explained the Flood,

It also left in northern countries the bodies of great quadrupeds, encased in ice and preserved with their skin, hair, and flesh down to our

own times. If they had not been frozen as soon as killed, petrification would have decomposed their carcasses. And, on the other hand, this continual frost did not previously occupy the places where the animals were seized by the ice, for they could not have existed in such a temperature. The animals were killed, therefore, at the same instant when glacial conditions overwhelmed the countries they inhabited. This development was sudden, not gradual...[Cuvier, Recherches sur les ossemens fosiles, I, 8-9, translated by Charles Coulston Gillispie]

Cuvier considered various explanations for such extinctions, initially considering the transient tidal wave Flood proposed by Pallas and Dolomieu which could explain the extinctions. But such transient causes could not explain the change in climate. Hoping to also account for earlier extinctions, Cuvier opted for DeLuc's explanation that supposed the earth had experienced regular revolutions between land and sea. As evidence of these earlier revolutions, Cuvier pointed to freshwater mollusks that alternated with marine shells which he supposed as having been deposited in earlier revolutions between land and sea.

Such floods caused by supposed earlier revolutions between land and sea diminished the uniqueness of the more dramatic 'Last Revolution' by which Cuvier explained the extinction of Europe's earlier mammals and which some were identifying with Noah's Flood. In contrast to DeLuc, Dolomieu, and Pallas, Cuvier had no religious motivation. Rather, he regarded the Bible’s account of Noah's Flood as but one several ancient testimonies to this worldwide Flood that occurred during the era of man. Moreover, Cuvier did not believe that whatever Flood was responsible for the megafauna extinctions had covered the earth's mountains. Though he contended that the Last Revolution occurred during the era of man, he suggested that at the time of the Flood, man had not spread outside the lands of the Bible.

I have already mentioned the archaeological problem caused by DeLuc, Cuvier, and those who followed the influential French scientist by explaining the Flood as being due to a revolution between land and ocean. It also caused geologists to suppose the lack of marine mollusks among these inland deposits as evidence against the Diluvium Flood. The true source of the waters of the Flood was put forth in 1808, when Cyprien Prosper Bard suggested that all these freshwater mollusks were the result of a single diluvian quake that mixed shells of various kinds. Deposits containing fresh-water shells would have covered rocks that contained marines shells due to the land having once been under the oceans.

Working as a biologist, Bard seems not to have seen this quake as also responsible for creating and spreading the erratics. If instead of catastrophes based on sea-to-land revolutions, the recent extinctions were the result of a single diluvian quake, Cuvier's 'Last Revolution' would indeed be unique and easily identified with Noah's Flood. Lamarck supported Bard. But Cuvier needed these earlier "Floods" to support his theory of extinctions against Lamarck's theory of evolution.

Due to troubles on the Continent caused by the Napoleonic wars, Cuvier's teachings first became known in the English-speaking world through the introduction and translation of his writings by Scottish geologist Robert Jameson (1774-1854) who was Regius Professor of Natural History at the University of Edinburg and founder of the Wernerian Natural History Society. Jameson's introduction and translation of Cuvier emphasized Cuvier's geology as confirming the recency of man as put forth in the Bible. Jameson also suggested that Cuvier's last revolution was possibly the same as the Noah's Flood in the Bible.

DeLuc and Cuvier's teaching of the ending of former worlds by earlier floods is responsible for the popular old earth explanation of Genesis put forth in 1814 by the Scottish Evangelical minister, Thomas Chalmers. According to Chalmers, the chaos mentioned in the second verse of Genesis resulted from the destruction of God's original Creation by an earlier Flood that is mentioned only in the first verse of Genesis. Though Cuvier's 'Last Revolution' may have been responsible for Noah's Flood, the earlier "Revolutions" could be seen as ending a Creation prior to the six-day creation mentioned in the first chapter of Genesis. In the first half of the twentieth century, this 'ruin and restoration' understanding, also called 'gap creation,' was the most popular understanding of Genesis held by conservative Christians, likely because it would eventually be embraced by the influential geologist William Buckland.

19

EQUATING THE DILUVIUM FLOOD WITH NOAH'S FLOOD

William Buckland (1784-1856), the first to write a full account of a fossil dinosaur, was a devoted disciple of Cuvier who would inherit his master's mantel to become the world's most influential geologist until the mantel would in turn be passed to Buckland's own student, Charles Lyell. With his friend William Conybeare (1787-1857), Buckland introduced the science of geology to Oxford University. Along with Cambridge University's Adam Sedgwick (1785-1863), these geologists who held positions at England's most prestigious universities introduced geologists to William Smith's (1769-1839) system of using fossils to identify the order of strata. Buckland's aim was to help his Oxford colleagues understand geology's connection to history and the Bible thereby increasing the prestige geology so as to introduce this new science to the Oxford curriculum.

Buckland pointed to the absence of fossils in the Primary Rocks as refuting Hutton's claim that the earth showed no evidence of its Creation. He pointed out that fossils of land animals such as elephants are not found in the earlier Secondary deposits but only in the later Tertiary layers. Contrary to Hutton, the fossils reveal that the earth has a progressive history of development as put forth in the Bible. Buckland became the leader of those who are referred to as *Progressive Creationists* to distinguish us from the near instantaneous creation taught by today's Young Earth Creationists.

Buckland was particularly interested in the surficial deposits that DeLuc and Cuvier saw as marking the divide between the present and former worlds. It was these most recent deposits that Buckland recognized the possibility of connecting geology to history as recorded in the Bible. Buckland popularized Cuvier's term *Alluvium* for the seasonal deposits from the flooding of present-day rivers which flow immediately above the *Diluvium*. But whereas the *Alluvium* contained the bones of living species, the *Diluvium* lying immediately below the Alluvium contained the bones of animals such as mammoths, rhinoceros, hyenas which if not altogether extinct are no longer present in Europe. The Diluvium also included the *erratics*. Buckland had observed smoothly polished round white pebbles made of hard quartzite in the vicinity of Oxford whose origin he traced to the vicinity of the English midlands. Supposing these to have been worn smooth in some ancient environment, Buckland suggested that a great Deluge had swept these *mini-erratics* southward far beyond the Thames River.

Buckland became famous for his study of caves as those in Bavaria where Cuvier had identified extinct bears. While some suggested that these bears had been swept into the caves by a great Flood, Buckland agreed with Cuvier that the bears were living in the caves when they were drowned by the Flood. Whereas Cuvier could reconstruct extinct animals, Buckland's skill lay in reconstructing extinct environments, which we now call *ecosystems*. The narratives and environments of what would later be understood as occurring during the Ice Age is due to Buckland's observations and imagination.

Buckland personally excavated the Kirkdale Cave in Yorkshire. Finding gnawed bones of animals ranging in size from mammoths, lions, tigers, and monkeys to water rats, he determined the cave to be an ancient hyena den. The bones of these extinct animals together with hyena coprolites lay in deposits between two layers of stalagmite. Here, Buckland also pioneered the scientific study of feces (coprolites). The upper layer of stalagmite served as a chronological marker dividing the

Fig 20 - Conybeare's fun drawing of his friend Buckland entering ancient hyena den of the Kirkdale Cave

deposits of the present age from the antediluvian world of extinct animals whose remains lay on an earlier deposit of stalagmite. Buckland equated these cave deposits with the extensive Diluvium found in river valleys. Thus, Buckland's hyena den served as a window into the vanished antediluvian world.

In 1820, Buckland published *Vindiciae Geologia; or the Connexion of Geology with Religion explained* in which he identified the Flood taught by DeLuc and Cuvier with the Noah's Flood of the Bible. As Buckland explained,

numberless phenomena have already been ascertained, which, without the admission of a recent and universal Deluge, it seems not easy, nay, utterly impossible to explain." [Buckland, Vindiciae Geologia, p38]

For most of the eighteen twenties, the Diluvium Flood served as the standard geological narrative. Geology was then seen as being in much harmony with the teachings of the Bible.

Whereas most professors limited themselves to books and museums, Buckland wore his academic gown on field trips to make it clear that such field research was part of the academy. He also traveled to geological sites on the Continent as he expanded and consolidated the arguments for the recent geological Flood as being the Flood of the Bible. His investigations benefited from Great Britain's expanding worldwide Empire and inspired scientific investigations as would soon include the young geologist and naturalist Charles Darwin added to the crew of the *HMS Beagle*. In 1826, the Governor of the East India trading base in Singapore but also a member of the Geological Society explored along the banks of a river in Southeast Asia. He shipped to England 7 chests of the bones of extinct animals of previously unidentified species of mastodons, hippopotamus, and rhinoceros which were no longer to be found in the region.

In 1828, a Captain Beechy sent bones to Buckland which he had identified along a river near Kotzebe Sound in Alaska that appeared similar to the extinct fauna that Buckland had found in the Diluvium. Buckland confirmed that they were similar to the species that he had found in the Kirkland Cave. Though of different species, bones of a similar recent extinction that had taken place in Australia had earlier been sent by the surveyor Thomas Livingstone Mitchell, who was also a member of the Geological Society to be examined by Robert Jamison and Georges Cuvier. It was thus becoming clear that the Diluvium Flood that caused the extinction of Europe's earlier wildlife had covered the entire earth.

20

Critics of the Diluvium Flood

While the Diluvium Flood served as the standard geological view during the 1820s, it was not without its critics including some who were not disciples of James Hutton. One such critic was the Scottish Presbyterian minister John Fleming (1785-1857). A disciple of Werner and Cuvier, Fleming did not like the way the Anglican William Buckland mixed geology and the Bible. Pointing to the olive leaf in the mouth of the dove that Noah sent out from the Ark, Fleming claimed that Noah's Flood was such a tranquil affair that it left the surface of the earth entirely undisturbed! While Cuvier and Buckland had supposed that at the time of the Flood mankind had not spread from the lands of the Bible, Fleming believed that Flood itself was limited to the lands of the Bible.

Though not a geologist, Fleming claimed that Diluvium was not always found under the Alluvium, but both types of deposits were found at various depths. This resulted in Adam Sedgewick coming to Buckland's defense by showing that the Alluvium did indeed overlay the Diluvium wherever they were found together. Fleming seems not to have had a problem with inventing facts to support his view.

Pointing out that wolves and bears were no longer to be found in England, Fleming attributed extinctions either to human hunters or slowly changing environments. Instead of studying extinctions from ancient times, Fleming insisted that they should be studied from historical times. Fleming ignored Cuvier and Buckland's observations that extinct species were found together in the Diluvium but were altogether missing in the Alluvium layers that lay above the Diluvium.

Fleming however pointed to the bones of an extinct Giant Elk (Fig 21) that had been found in peat showed evidence of having been killed by a human hunter. Whether the bones of this extinct species indicate that the peat had been deposited by the Flood, this had to carry great weight with Buckland who rejected the view that humans were living in Britain at the time of the Flood.

The other major critic of the Diluvium Flood was the amateur geologist George Poulett Scrope (1797-1876) who resented the way Buckland's Diluvium Flood stopped further inquiry. Scrope was referring to how the Diluvium Flood was becoming the accepted geological view. He was himself manifest evidence of how false his concern. Scrope however would be instrumental in shifting geology to a way that did stop further inquiry.

Fig 21 – Extinct Giant Elk

Unlike Hutton, Scrope held to a directionalist view of the earth's history. But he was no less convinced that the U-shaped valleys then being explained by the Flood could be explained by ordinary erosion should one allow unlimited amounts of time as Hutton had suggested. Scrope pointed to the steep and wandering valleys through which flowed the Moselle and Meuse Rivers as unlikely to have been formed as the result of a single Flood. However, neither was this evidence that a Flood did not scour though these pre-existing narrow valleys at such a rate as to fail to deposit Diluvium as occurred in the valleys. But Scrope's argument against the Diluvium benefitted from the excessive erosion that many Diluvialists including William Buckland attributed to the Flood, perhaps including the volcanic cones that Scrope chose to study.

To support his claim that ordinary erosion was sufficient to explain what was then being attributed to the Diluvium Flood, Scrope studied the volcanoes of central France. Earlier geologists believed that the cones of the volcanoes supposed as having erupted before the Flood had been eroded away by the Flood while volcanoes erupting after the Flood had intact cones. Scrope showed how basaltic flows from volcanoes supposed as having erupted from before the Flood had altered the direction of streams created by basalt flows from volcanoes supposed as having erupted after the Flood, thus seeming to eliminate one argument for the Diluvium Flood.

The most ambitious critic of the Diluvium Flood was Buckland's own student, the wealthy and well-connected young lawyer Charles Lyell (1797-1875). Lyell agreed with Scrope that the Diluvium Flood licensed the use of the Bible in geology due to being taught at England's two leading universities. Resenting the prestige of England's premier universities where only ordained Anglican ministers could teach, Lyell sought a secularization of England's universities. Though coming from a family of Tories, he sympathized with the Reform movement which would open England's universities to non-Anglicans and to non-ordained professors.

Though nominally a Christian, Lyell aimed to entirely separate science from the Bible and history as had also been the aim of James Hutton. He would succeed in his aim of changing what was formerly called *natural theology* to the purely secular endeavor that would come to be known as *science*. Lyell suggested that the biblical Flood, which for him was a matter of interest for theologians, had occurred in the region of the Caspian Sea or perhaps in the Jordan Valley. Such a local and ephemeral phenomenon would thus have no consequence for the science of geology.

Lyell endorsed Scrope's view of the slow excavation of valleys. Two French naturalist Croizet and Jobert had identified species associated with the Diluvium extinctions beneath the basalt flows studied by Scrope. Combining Scrope's theory of the slow excavation of valleys with Fleming's theory of the gradual extinction of species, Lyell supposed that he had pushed back the date of the Diluvium far before the era of man. However, Lyell was basing this on the contingent and hardly uniform phenomenon of volcanic eruptions.

21

How Charles Lyell Re-envisioned the Diluvium as the Deposits of Drifting Icebergs

After his name became known throughout the world from his three-volume *Principles of Geology* published between 1830-1833, Charles Lyell became more famous than Buckland and Sedgwick. His success was due (1) to his excellent writing style; (2) to advocating for his new theory in the manner of a skilled attorney; (3) to his prestigious scientific connections including an indulgent and friendly scientific book publisher; (4) but above all to the fact that, during a time of rapid progress in the physical sciences, many learned theologians including the influential Rev William Whewell (1794-1866) were hoping to see a scientific geology replace the old way of doing geology from the light of the Bible! None can better remove the light of the Bible than ordained Christian theologians.

In his *Principles of Geology*, Lyell championed James Hutton's uniformitarian understanding of the earth. Whewell coined the terms *uniformitarian* and *catastrophist* as well as helping Lyell coin new geological periods. It would be from this celebrated book that most would learn of an ancient earth which many Creationists wrongly suppose as having been the result of Lyell's uniformitarian geology. Rather the catastrophist geology of Progressive Creationists such as DeLuc, Cuvier, and Buckland that Lyell challenged in this book established the fact of an ancient earth. Lyell however did use his book to vastly stretch the antiquity of the earth while adding several new periods to the earth's more recent geological past. Because of Lyell, most do not know that belief in an ancient earth came from studying the earth in the light of the Bible.

Most important, it was this book that convinced even most catastrophists that the Flood of Noah was nothing more than a geographically local Flood. Of even greater importance, is that most Christians would no longer hear of the Diluvium Flood, the evidence of which Lyell would explain as but another long geological period.

LYELL'S UNIFORMITARIAN GEOLOGY

Lyell's chief obstacle in challenging the Diluvium Flood was explaining the placement of the great *erratic* stones. For this task, he modified the iceberg drift theory espoused by Hutton's close associate James Hall. But instead of using Hall's theory of diluvian waves, Lyell employed DeLuc and Cuvier's explanation of shifting elevations between land and oceans. But rather than using DeLuc and Cuvier's sudden revolutions, Lyell supposed the inundations of the oceans to have been gradual and regular.

FIG 22 - Frontispiece to Lyell's *Principles of Geology*

Lyell illustrated his new theory by referring to the mollusk encrustations on the Temple of Serapis in the Bay of Naples which served as the frontispiece to his *Principles of Geology* (Fig 22). This drawing showed that the Mediterranean Sea had risen to the level of the mollusk encrustations before falling to the current level of this sea at the time of this line drawing.

Current geologists explain this anomaly as due to the frequent earthquakes which alternate with elevation of land due to Mount Vesuvius, which destroyed nearby Pompeii and Heracleum. Note that to explain Diluvium deposits as having been gradual and local, Lyell used a truly exceptional local phenomena as the universal mechanism for explaining geological strata and wildlife all over the earth for the entire period of the earth's history! This is hardly inductive reasoning base on empirical science implicit in the subtitle of his book, *'Being an Attempt to Explain the Former Changes in the Earth's Surface by Reference to Causes Now in Operation.'* Rather, his subtitle should have been, *'Being an Attempt to Explain the Former Changes in the Earth's Surface by Universalizing an Exceptional Cause Found at But One Place in the Earth.*

As did Hutton, Lyell denied that the earth had ever experienced any development or universal change. According to him, aside from local changes the earth has essentially always been as it exists today. In addition to following Hutton in using erosion and volcanoes to explain the earth's geological cycles, Lyell postulated slow changes in climate caused by yoyo movements in the elevation of continents, to which he added fluctuations due to changes in ocean currents such as the Gulf Stream. Slow changes in local fauna were supposed as being due to these slow changes in climate.

AWFUL CHANGES.

MAN FOUND ONLY IN A FOSSIL STATE——REAPPEARANCE OF ICHTHYOSAURI.

A Lecture.—" You will at once perceive," continued PROFESSOR ICHTHYOSAURUS, "that the skull before us belonged to some of the lower order of animals ; the teeth are very insignificant, the power of the jaws trifling, and altogether it seems wonderful how the creature could have procured food."

Fig 23 – De la Bech's cartoon depicting Lyell as Professor Dinosaur lecturing about a human skull

This meant denying the universal extinctions and the appearance of any new species for the earth as a whole! As illustrated in De la Beche's famous cartoon showing Professor Dinosaur lecturing over a human skull (Fig 23), Lyell believed that dinosaurs were still to be found someplace on the unexplored earth and might one day reappear during a time when humans had become extinct in England!

CHANGING DILUVIUM DEPOSITS TO ICEBERG DRIFT

According to Lyell, the great *erratic* stones together with the various types of gravel and soil in the Diluvium came from the bottom of frozen oceans. Lyell attributed the frozen oceans to the rise in the elevation of the continents and to shifting currents such as the Gulf Stream that currently warms the British Isles. As the oceans slowly rise and the coastal land sinks as happened in the Bay of Naples, Lyell claimed that icebergs containing the Diluvium drifted upon land where they unceremoniously dump the *erratics* together with the various gravels and soils that had gotten frozen into the icebergs. Thus did Lyell renamed the Diluvium deposits *Drift*.

The Drift (i.e. the Diluvium) however was not just found in low-lying areas, but also in mountains where the *erratics* had been famously studied. Lyell claimed that these mountain *erratics* had been carried on frozen ice rafts in ancient lakes that once formed high in the Alps. According to Lyell, these great rocks were also responsible for shaping valleys and digging lakes, including the Great Lakes of America which lie just above the Diluvium!

This is a truly fantastic theory! It is no less fantastic than the Flood theories of today's Young Earth Creationists but supposed based on what we are assured is as a sound scientific explanation based on the uniform behavior of the earth 'by reference to causes now in operation.' Where has anyone seen such a thing? Imagine large rocks carried by icebergs sculpting what would become Lake Superior! Even if one can suppose a rock jutting as a long spade from the bottom of an iceberg digging out Lake Superior, how long before the earth's pull against the rock causes the melting iceberg to loosen and drop the spade rock? This becomes all the more the case due to the long periods required by Lyell's gradualist theory. Did Lyell even know how deep is Lake Superior?

One can understand why I compare such theories now embraced by mainstream geology to the unconstrained imaginations of Creationists as former school attempts to explain away and the latter to explain the

Flood. Such silliness continues to be embraced in the Ice Age theory that would soon replace Lyell's iceberg drift theory. Grown men imagine, believe, and teach such things! The irony is that this became Science's first great victory over the Bible, leading to Science replacing the Bible as the modern world's intellectual authority! Such great irony should leave us breathless!

One can however understand why, aside from the larger public who fail to understand Lyell's theory, Lyell had trouble convincing professional geologists except the young Charles Darwin who was delighted to read Lyell's just published book during his voyage on the famous *HMS Beagle.* Darwin enthusiastically confessed that he learned to interpret everything he observed through the lens of Lyell's new paradigm. He leaned from Lyell that one cannot suppose the fossil record provides an accurate picture of the earth's history due to what they claim as the extreme rarity of preservation. Goodbye inductive science! Find or create what appears as a rare exception and build on that foundation whatever science one pleases! Lyell's explanation of why we do not find human remains among the dinosaurs and earlier forms of life is due to the rarity of preservation. However, this is not so different from how Young Earth Creationists explain the failure to find the remains of the antediluvian world. In basing things on rare contingent causes, Young Earth Creationism is a mirror to Lyell's uniformitarian science,

As his critics pointed out, Lyell's system requires that *a priori* dogma rule, regardless empirical discoveries. He made the same argument against the record of geological strata. According to Lyell, there was no historical development but only chance preservation of effectively eternal forms of rocks and lifeforms that had not changed since earth's Creation, the origins of which should not be the concern of geologists. Thus did Lyell argue in the most extended sections of the first volume of his *Principles of Geology.*

James Hutton had likewise argued against trying to discover the origins of the earth. As did Hutton, Lyell denied that the earth as a whole had ever changed. Using a title which suggested that he was but extending Isaac Newton's *Principles of Natural Philosophy* to geological science, Lyell believed that he had discovered a scientific mechanism for all the earth's ongoing changes. However, in his *Origin of Species,* Darwin would reject Lyell's uniformitarian science because any explanation of origins must violate a strictly uniformitarian explanation. This created no small tension between Lyell and his student. But Lyell believed that his *Principles* had dealt the death blow to any theory of evolution! Even so, Lyell's book inspired theories of evolution for the very reason that he explicitly refused to give an explanation for the origins of anything, including species!

REPLACING THE TERTIARY WITH MULTIPLE RECENT GEOLOGICAL PERIODS

Contrary to what has been supposed, Lyell sought to eliminate all historical change in geological and biological forms whether due to Creation or evolution. Why this has been misunderstood is due to Lyell's role in introducing the geological column to the large public and by having invented long geological periods within the Tertiary, which he replaced with new epochs named *Eocene, Miocene,* and *Older Pliocene*. He also renamed the Diluvium as *Newer Pliocene* in order for it to appear but a brief phrase in his artifically expanded timescale. His expansion of the Tertiary period was to make geological changes seem gradual and continuous with the Diluvium which he had earlier named *Pleistocene,* the name that would eventually prevail for the deposits of the Flood.

This was no exercise in empirical science but an exercise in fitting any and all data into his theoretical dogma. Lyell's "empiricism" was due to publishing his many interrogations of naturalists who contemptuously referred to him as 'the Pump' for taking their findings and running them through his uniformitarian interpretation machine. Lyell used his famous name and connections to intimidate naturalists into cooperating with him.

DETERMINING AGE OF FOSSIL BEDS BY PERCENTAGE OF LIVING SPECIES

Lyell's way of ordering geological deposits was according to the percentage of living mollusks by which he dated various Tertiary basins. The great changes in species that occur during the Tertiary concern land animals, especially mammals which appear in this period. Thus, if one uses fossils as index for geological ages, in the case of the Tertiary, he ought to have used the fossils of mammals! Mollusks however are among the oldest of fossils, appearing from the Cambrian Explosion. By contrast, mollusk extinctions are highly debated even in the Cambrian period where the variations are greatest. There is also great variation among living mollusks. The paleontologist Daniel Prothero tells us that most of the Paris Basin mollusks are restricted to that region and cannot be recognized elsewhere, so it is impossible to correlate with any other region.[16]

USING A RECENT VOLCANO TO SUPPOSE A LONG AND ANCIENT PLEISTOCENCE.

Lyell's chief example of using the percentage of current mollusks species was instrumental in his demonstration of the enormous time represented in the geological periods. Not surprisingly he employed volcanoes, the most unpredictable of all geological processes, specifically Mount Etna in Sicily which is erupting again even as I write this on July 6, 2024. Mount Etna's eruptions produce new flows of lava which cool and become basalt. Beneath the lowest basalt flow from Mount Etna, Lyell found evidence of the *Newer Pliocene* associated with mollusks similar to those still living in the Mediterranean Sea. From what he learned about historical eruptions of Mount Etna, Lyell guessed the most recent Tertiary deposits (i.e, the Diluvium) date from at least several hundred thousands of years. This was Lyell's most impressive example of creating a supposed great passage of geological time for what was in fact but a but a short period in geological history.

Lyell's argument of course depended on his calculations of the basalt flows of Mount Etna. Von Buch (1744-1853), who I earlier mentioned as calculating the great velocities of the erratics, determined that such lava flows as Lyell uses can only happen at slope angles less than 3 degrees. According to Von Buch, the steep slopes of Mount Etna point to this volcano as having been created as a result of 'great displays of power from within' the body of the earth. The Diluvium beneath the lowest flows of this mountain point to Mount Etna as postdating the Flood, likely having been created by the changes in the earth due to the Flood.

If Lyell was looking for a way of creating new periods to make the Diluvium deposits seem but an ordinary geological strata, supposed percentages of living mollusks was an excellent way of doing it. In his defense, Lyell did not invent this method which dates back to vertebrate expert Georges Cuvier who used freshwater shellfish to define the entire range of geological periods based on fossils of invertebrate! As I mentioned in Chapter 18, as early as 1808 Cyprian Prosper Brard suggested that these freshwater mollusks were the result of a great

---

[16] Donald Prothero, *The Eocene-Oligocene Transition: Paradise Loss,* p. 42, quoted from Tony Edger's blog

diluvial quake that had mixed shells of various kinds. Thus, Cuvier's "last" Flood was sufficient to have mixed these freshwater mollusks with the more ancient fossils. Unfortunately, Cuvier needed his earlier "revolutions" to defend extinctions against Lamarck, the expert on invertebrate who supported Brard. Brard also pointed out that mollusks continuously adapt to changing salinity. Thus, they can be no reliable marker of time. In Chapter 59, I will explain just how profoundly correct was Brard's suggestion.

While the dumping of the Diluvium by icebergs explained the lack of a simple bedding pattern, it could hardly explain why Diluvium is never found in the deeper strata of the earth. The same was true of human skulls. Lyell explained both as being due to the rarity of preservation. He used the depth of Drift (i.e. the Diluvium) to estimate the duration of what he supposed were gradual and uniform deposits, guessing it to be several hundred thousand years. Such calculations ignore the fact that the Diluvium is found at various depths, depending on the topsoil and depth of base rock in various parts of the earth.

Such gradual flooding could not have been the same as Noah's Flood which however was precisely Lyell's aim. Lyell allowed that Noah's Flood might have been a local flood limited to region of the Caspian Sea or perhaps the Jordan Valley. Though Lyell's uniformitarian geology would be defeated by the triumph of an Ice Age and by his disciple Charles Darwin's theory of evolution, Lyell was consoled by the fact that he had succeeded in his chief aim which in his own words was 'freeing geology from Moses.'

22

The War Waged by Professional Geologists Against the Scriptural Geologists

Deists such as Charles Lyell referred to the Genesis account as 'Moses' rather than being that of the much loved Bible long championed by Protestants. By referring to the account as "Moses," rationalists could dismiss Genesis as the account of the despised rabbinical Jews with whom they were acquainted. German scholars in Lyell's circle were envisioning the ancient Jews as a minor tribal nation on the fringes of the ancient world and a people who believed they could be saved through their legalist religion. They were however projecting the practices of the rabbis from whom they learned Hebrew to the Jews of the Bible. This became an essential element in the German higher criticism of the Bible.

In contrast to the French and German-speaking universities where science had become fully secularized, Lyell was embarrassed that in the English-speaking world the Bible's account of origins continued to be used for understanding geology. This was because most universities in the English-speaking world were still in the hands of churches. Most embarrassing for Lyell was the fact that the leading geologists of the Diluvium Flood were ordained ministers in the prestigious English universities of Oxford and Cambridge. Though coming from the elite Tory class who refused to allow non-Anglicans to attend these universities, Lyell championed the creation of secular universities including King's College in London where he temporarily accepted a position as professor of geology.

With the assistance of Whewell, Lyell promoted the notion of a professional 'man of science' who would replace the authority of religious ministers regarding matters of science. Such professional geologist would be members of the London Geological Society, which sought its independence from the Royal Society whose founding members were ordained Christian ministers.

Most disliked by these men were those called *scriptural geologists* who, as their name indicates, insisted on using the light of Scripture to understand geology. This of course is how I do study geology and all sciences. Though scriptural geologists rejected the Baconian compromise with Scripture, they were diverse in their origins and expertise, many being members of the free churches who were generally more conservative than those from Oxford and Cambridge. The writings of scriptural geologists such as Andre Ire (1778-1837) and Granville Penn (1761-1844), grandson of William Penn, were popular with the public.

Moreover, geology was the most democratic of the sciences, open to any who wished to collect and study minerals and fossils or report their field observations of rock formations, thing valuable to the new science. Thus, Lyell had two complementary aims: (1) freeing geology from the Bible and (2) turning geology into a professional science which could exclude the scriptural geologists. However, his chief obstacle in removing the light of the Bible from geology were the Diluvium geologists at Oxford and Cambridge such as William Buckland and Adam Sedgwick from whom Lyell had himself learned geology. Fortunately, as did Whewell, Buckland and Sedgwick were joining his effort to exclude the scriptural geologists from professional geology.

Adam Sedgwick was the first of the leading Diluvialists to abandon the worldwide Flood, agreeing that the evidence of the Diluvium Flood existed only in certain localities. In 1829, he used his last address as President of the Geological Society to recant his former support of a worldwide Flood, his chief reason being the failure to find the evidence of humans in the Diluvium. Though not uncritical of Lyell and continuing to believe that the evidence of many localities was best explained by the Diluvium Flood, Sedgwick encouraged and promoted Lyell's gradualist approach to geological phenomena. However Buckland and Conybeare continued to defend the Diluvium Flood.

Even Buckland would eventually acknowledge that the Diluvium Flood was too early to be identified with the Flood of Noah, thus undermining his original aim to connecting geology with the Bible. Buckland's reason for abandoning his claim that the Diluvium Flood was Noah’s Flood was the same as that of Sedgwick, their failure to find human remains in the Diluvium. In fact it was due to his refusal to acknowledge the evidence of human remains which he had himself found in the Diluvium.

Likely, William Buckland's acknowledgment that the Diluvium Flood was too early to have been Noah's Flood was his quiet way of disassociating himself from the scriptural geologists, some of whom accused him of heterodoxy. That could impact his appointment to a prestigious bishopric. Though most scriptural geologists were acquainted with the developments in geology, made empirical contributions to this new science, and a few were members of the geological societies, due to insisting on using the light of the Bible for geology, they would soon be marginalized in learned circles and pushed from professional geology.

Lyell's way of freeing geology from Moses was to write to the learned public, especially the conservative audience who read the Tory's *Quarterly Review.* Lyell began by introducing the public to the new science of geology while undermining the Diluvium Flood as when he reviewed Scrope's book on how erosion could explain valleys previously believed to have been sculpted by the Flood. It was however the great success of Lyell's *Principles of Geology* that allowed him to exclude the scriptural geologists from the professional geology.

Because extinct fossils were just beginning to reveal the earth's great antiquity, most scriptural geologists continued to hold the young earth view that had informed geology from the era of the Scientific Revolution. Many of the most popular scriptural geologists did fail to understand the Diluvium Flood. However, one should not equate the broader arguments of the scriptural geologists with today's Young Earth Creationists whose roots are in the *New Geology* of Seventh Day Adventists Ellen G. White and George McCready Price as discussed in Chapter 3 above. A few scriptural geologists also taught an ancient earth, but all of them believed that Noah's Flood was worldwide.

Thus, did Lyell create what by the end of the nineteenth century would be seen as the conflict between Science and the Bible which, as also the case of the theory of evolution, predated Darwin. The word *science* began replacing what was formerly called *natural theology* or natural *philosophy.* Naturalists such as Lyell referred to themsleves as *men of science.* The word *scientist* was adopted in America near the beginning of the twentieth century though only later in Europe. By then, the conflict would concern Darwin's theory of evolution. Against a literal reading of the Bible, Science would be on the side of evolution which won the war in influential American public opinion after the famous 1925 Scopes Trial. Unfortunately, the Diluvium Flood was soon forgotten due to the sensation of the new theory of the Ice Age, a new way of explaining the evidence that had led geologists to their discover of the Diluvium Flood.

23

IMAGINING AN ICE AGE TO EXPLAIN THE DILUVIUM DEPOSITS

Fig 24 - Agassiz's sketch of an erratic sitting atop a glacier

In the meantime, a new theory emerged to explain the Diluvium deposits that Lyell and Darwin were attributing to work of drifting icebergs. If the displacement of the *erratics* had long been attributed to a liquid Flood, perhaps they could be explained by a slow-moving frozen Flood as happens when after years of deep snow, glaciers may slowly flow from snow-capped mountains like Mount Ararat. Such a theory developed in Switzerland where existed both *erratic* stones and mountain glaciers.

The popularization of this theory to explain the worldwide deposits of what at the time were understood as either *Diluvium* or *Drift* was the work of the world's leading expert on fossil fish. Louis Agassiz (1807-1883) was already a famous naturalist when he used his reputation to promote the theory of an Ice Age. Agassiz had learned a theory proposed by local Swiss geologists that glaciers had once extended far down the Alps where are found drumlins and *erratics*. Though initially rejecting this old theory once proposed by Hutton,

Agassiz convinced himself that much could be explained by supposing that glaciers had once covered much of Europe.

Aggasiz's famous name served as a platform for what otherwise would have probably remained an obscure theory. He presented this idea to the 1837 Swiss Society of Natural Sciences where it received a completely negative reception. The great Alexander Humboldt advised Agassiz to return to studying fossil fishes. Leopold Von Buch, who had studied the dynamics of erratic stones, sharply criticized the possibility that glaciers could ever explain the displacement of great erratics. The supposed erratic sitting atop the glacier in his line drawing (Fig 24) would have tumbled from the mountain as we see on our climbs of Mount Ararat. Aside from sharing the same slope, the movement of such rocks will be largely independent of the ice glacier.

Agassiz presented his idea in a memoir outlining the historical evidence for oscillating glaciers in Switzerland and proposed that vast ice sheets had once covered much of Europe. Agassiz's big break was due to his close friendship with William Buckland, still one of the world's most respected geologist after abandoning the Diluvium as having been Noah's Flood. This happened in 1840 when Agassiz accepted Buckland's invitation to visited Britain and read a paper to the British Association of the Advancement of Science. Buckland believed that he had converted Lyell to support the Ice Age, though that was premature. What most impressed Buckland was that parallel grooves along bedrock could be better explained by glaciers sliding large *erratics* across bedrock than could Lyell's iceberg drift theory.

Buckland's support was key. Agassiz and Buckland also assumed the same kind of Ice Age glaciers were responsible for the Diluvium that the latter had found in the tropics of southern Asia. In 1839, Agassiz claimed that the Ice Age had extended to the Mediterranean and thus had covered Egypt! By 1846, when Agassiz came to America to take a position at Harvard University, the Ice Age was all the rage in Britain and America. Agassiz embarked on a trip to Brazil where he discovered evidence of glacier deposits in the Amazon jungles! As Buckland had found evidence of the diluvium in the jungles of Southeast Asia and in the highlands of the Himalayan Mountains, Agassiz concluded that glaciers had once covered the entire earth. Not surprisingly, Agassiz's teaching of what was being called *Snowball Earth* had the effect of cooling enthusiasm for an Ice Age. Thus, Lyell and Darwin continued to prefer their Drift theory to such a non-uniformitarian theory as a recent great Ice Age.

In 1852, a scientific expedition to Greenland discovered that great ice sheets covered most of this large island. A few years later, ice sheets were discovered in Antarctica. As southern limits to northern ice sheets became accepted, so was the Ice Age. The Diluvium does appear somewhat different in northern climates such as Scotland and Canada where less vegetation grows and where exists less topsoil by which to cover bedrock, gravel, and *erratics*. The attention given to explaining an Ice Age now restricted to the colder northern regions deflects attention from the fact that the Diluvium deposits including *erratics* are no less to be found in the earth's temperate and tropic regions. The latter is due to the *erratics* having been deposited by the worldwide Flood of Noah.

24

Calculating the Time Since the Flood (or Since the Ice Age)

I return to the Diluvium Flood and how American geologists found a way of dating the Flood. I have waited to cover how geologists dated the Flood because it would also be used to date the time since the end of the Ice Age, also to point out Lyell's role in vastly expanding this era so as to make it easier to explain these phenomena by causes still in operation.

Inspired by two of DeLuc's observations: (1) that rivers date from the time of the Flood and (2) that silt lying above ancient monuments under river deltas might serve as a measure of time, Dolomieu traveled with Napoleon's troops to Egypt where he hoped to find a monument beneath the Nile Delta by which he might use to calculate the time since man first appeared in Egypt. Dolomieu's project ended with the downfall of Napoleon, but others continued this work as published in the famous *Description de lEgypte* between 1809 and 1829.

Based on the 72-foot depth of a monument found in the delta and his estimated mean rate of sediment build up in the delta of 2.5 inches per centuries, M. Rosiere calculated the time since the placement of the monument of to be about 30,000 years. That is at least 25,000 years too early, but no one then had any better method of scientifically determining the beginnings of Egyptian civilization. At a time when there existed few ways of calculating the age of the earth, some supposed this as being the age of the earth, at least challenging the 6,000 years which many supposed as the age of the earth.

American geologists however believed they had a sound geological method of determining the time since the Flood deposited the Diluvium. As everyone acknowledged, the American Great Lakes rests on Diluvium deposits. The erosion of Niagara Falls by the Saint Lawrence

River had begun at Lake Ontario. Knowing the time since the Flood was a matter of estimating the rate of erosion of Niagara Falls and multiplying that by the distance of the Falls from Lake Ontario. In 1829, the British geologist Robert Bakewell used historical information from the people who lived near Niagara Falls to estimate the rate of erosion. Based on estimated erosion of 3 feet per year, Bakewell calculated that it had been about 10,000 years since Niagara Falls began eroding from Lake Ontario. The sensation this cause in America was due to having demonstrated the age of the earth as being earlier than about 6,000 years as most Americans then believed. Diluvialists however could use this as a measure of the era since the Diluvium Flood.

Recognizing the importance of these measurements, Lyell himself twice visited the Falls in 1840. To expand the time that had elapsed since what he was now calling *Drift,* Lyell guessed the erosion to be only 1 foot per year to calculate at least 30,000 years. Lyell was no doubt acquainted with the calculations published in the famous *Description de lEgypte!* Keep this in mind for understanding the "scientific" methods of the man most responsible for shifting the authority recognized by European Christians from the Bible to Science and wresting the sciences from Christian universities to become secular professions.

However appreciative that this famous European geologist took note of geology in America, American geologists such as Yale's James Dana did not like Lyell so cavalierly dismissing their careful work. Strong opposition from American geologists who had an excellent empirical basis for their measurements, restrained Lyell from expanding the Holocene, the time coming to be known as 'the end of the Ice Age,' to 30,000 years.

American geologists continued to work on refining their calculations based on the erosion of Niagara Falls to near the beginning of the twentieth century. In 1878, Alexander Winchell calculated the rate of erosion for St. Anthony's Falls near the beginning of the Mississippi River and found that was by then being called the Ice Age must have ended only about 8,000 years ago. In Chapter 27, I will explain how Lyell agreed to compromise the time since the end of the Ice Age as being 12,000 years ago and the most interesting basis of that determination which would be based on the beginning of Egyptian civilization that could be no earlier than the end of the Ice Age.

25

WHY GLACIERS CANNOT PLOW THE EARTH OR MOVE LARGE ROCKS

The ability of Ice Age enthusiasts and glacier scientists to ignore the massive evidence of their supposed Ice Age in temperate and tropical region is hardly less jaw jarring than their ability to believe the fanciful explanation of how these great rocks were transported by glaciers in ways that are so contrary to the principles of materials science, even to the basic laws of physics such as gravity. Lyell however had already trained geologists to ignore the principles of physics in his *Principles of Geology*.

In line with Lyell's iceberg drift theory, glacier scientists tell us that rocks freezing to the bottoms of glaciers operate as a great file, scraping and transporting the earth below them, thus leaving behind what had previously been understood as deposits from the Diluvium Flood.[17] They also point to what they call terminal moraines, some of which lie in front of the snouts of downhill glaciers.[18] There are also lateral moraines along the sides of the glaciers.

As I learn from the climbers and my own experience with the glaciers on Mount Ararat, rocks do freeze to the bottom of glaciers. But the tensile strength of this part of the glacier is the greatest of any part of the glacier. Instead of scraping the ground, the frozen rocks just retard whatever downward flow might occur. Due to gravity and the plasticity of ice, glaciers of sufficient depth do slowly flow downhill, but above the frozen bottom until the bottom of the glacier turns to water. The water reveals that the glacier is melting and disappearing along its downward slope but it is not plowing rocks! This is why we climb Mount Ararat

---

[17] Imbrie and Imbrie, *Ice Ages, Solving the Mystery* (1979, Harvard University Press), p. 48
[18] Doug Macdougall, Frozen Earth: The Once and Future Story of Ice Ages (2004, University of California Press) p. 49

while the rocks are still frozen and thus falling rocks do not so endanger us.

The geologist Sir Henry Howorth (1842-1923) was appalled by the notion that ice sheets could defy the laws of gravity by flowing uphill, even over mountains, digging out deep lakes by dragging great boulders. It was however scarcely less difficult to believe than Lyell and Darwin's equally fantastic Drift theory. That geologists could be convinced of an Ice Age was Howorth's 'Glacier Nightmare.'[19] However, even Howorth accepted glacier explanations for the rock deposits beneath ice-capped mountains.

If not a dream, there was a feat of great imagination behind drawings of a miles-high wall of ice sliding towards what is now New York City. Though much lower, such ice cliffs as can be seen in Greenland or Antarctica are created when icebergs fall into the ocean, melt or float away. But where in the interior of continents might they fall into the ocean? Large glaciers do move towards the ocean through sloping valleys in the same way that unfrozen rivers flow towards the ocean. But to imagine a lateral force propelling icesheets across flat land, even uphill(!), glacier scientists must imagine snow somehow falling only in selected spots from which these icesheets are somehow supposed to spread. That only happens in Antarctica and Greenland where glaciers reach the coasts.

As one travels toward the Arctic regions, a permanent ice covering builds up slowly and gradually in the frozen tundra and melts in the same slow way but without dragging rocks aside from what might flow from meltwaters. Agassiz's Ice Age theory however was not inspired by ice sheets covering flat lands, but by observations of *erratics* and deposits found in the valleys of mountain glaciers.

I had an opportunity to view how these "glacier" rocks tumble from ice-capped Mount Ararat. As the case of high mountains throughout the world, Mount Ararat has been slowly growing in height causing rocks to regularly fall from its higher levels. Rockslides frequently occur even in the far lower Appalachian mountains. But these rocks endanger visits to the site of the Ark. The winter snows cause the mountain's glaciers to grow and freeze to the rocks beneath them. During summer melts, dams of water gather under these glaciers often forming small lakes. In 2022, my friend Parasut sent me a video that he recorded of a rapidly gushing stream of water shooting down the mountain from inside the

---

[19] Sir Henry H. Howorth, *The Glacial Nightmare and the Flood* (Vol I, 1892; Vol II, 1893)

glacier that covers the Ark. The rushing stream of water was carrying rocks, large and small. It is gravity, not glaciers that is responsible for "Ice Age" deposits, even those just below ice-capped mountains!

Fig 25 - The author standing near an "Ice Age" erratic in Charlotte, NC

To understand the blinkered way that glacier scientists follow the established Ice Age narrative, just consider the "glacier" erratic by which I stand in the Big Rock Nature Preserve in my hometown of Charlotte, NC which is not far from where the Catawba River crosses the North and South Carolina border (Fig 25). The eroded soil which was long ago washed into the Catawba River helped reveal these rocks whose origins are from the granite of the bedrock of the North Carolina Piedmont. My son Paul may have found the origins of these rocks near Huntersville, NC, which is in northern part of the same county where these *erratics* are found. There they exist in great piles, as if most of the rocks having fallen to the sides at the place of explosion of the waters of the Great Deep.

We also found on some of these rocks, the famous scratches believed caused by rapidly moving rocks. How these erratics got to my hometown of Charlotte cannot be explained by any known geological process, especially since Ice Age glaciers are not supposed to have covered this part of North Carolina. I have discussed these rocks with some glacier scientists. They have been as perplexed as me.

Were the glaciers that moved these rocks equipped with a snowplow? During a sufficiently long snowy and cold climate as supposed for an Ice Age, a glacier might slowly cover a rock this size. However, along hilly ground as in this part of North Carolina, a glacier will not even move, much less push such a heavy rock. Whatever deposited these *erratics* as also Mecklenburg County's famous red clay which is mixed with all manner of sandstone, it could not have been glaciers. It could only have been great amounts of swiftly moving water, the sources of which I will explain in in Chapter 59.

Interestingly, Mecklenburg County is named for the same German state where DeLuc pondered the Stone Age house covered by the deposits of Noah's Flood. This is because the citizens of Charlotte named their county to honor the home of DeLuc's patroness Queen Charlotte, wife of King George III whom the citizens of Mecklenburg Country regarded as a tyrant and from whom they declared their independence on May 20, 1775!

Part IV

Imagining and Dating the Antiquity of Man

26

The Biblical Roots of Historical Understanding

The chief contention of the Deists James Hutton and Charles Lyell is that science, which aims to describe the eternally fixed nature of the world, should have no history. While science is rightly described as natural philosophy, according to these Deists what we currently know as *natural history* should not exist because *Nature* has no history. Natural history is also something that Aristotle, the greatest of the ancient scientists, would have rejected. Both Hutton and Lyell rejected the search for origins as being a scientific endeavor. Likewise, both men rejected an event such as the Flood that would have given the earth a geological history. That we currently recognize that the earth and universe do have a history is due to influence of the Bible on the sciences.

While the pagan Greeks are justly regarded as the fathers of philosophy from which such eternalist views of the sciences are based, historical understanding requires that people value, record, and preserve the memory of the people, times, and events of the past. The Romanian historian of religion Mircea Eliade (1907-1986) noted how the Western world's linear sense of history is grounded in the Bible. Because history as known from the Bible has a beginning and direction, I replace Eliade's familiar phrase 'linear sense of history' with 'the arrow of history.'

The Bible's arrow of history follows world-changing miracles and events from the successive acts of Creation in Genesis to the events of the Flood and Exodus, together with history-changing persons such as Abraham, Moses, David, Jesus, Constantine, Mohammed, and Martin Luther. Due to being based on contingent acts and people and thus not subject to discoverable laws, history can only be known from reliably preserved human testimony and perhaps divine revelation concerning things not witnessed by humans.

A distinction must be made between the *historical,* which is based on contingent (unpredictable) causes, and the temporal dimension belonging to the *historicism* of philosophical, theological, ideological, or scientific systems. The temporal dimension of *historicist* ways of understanding is determined by philosophical or theological determined events. For example, there is a determinist arrow of time (though not an arrow of history) belonging to the Neoplatonist philosophy of emergence that Augustine of Hippo employed in his famous *City of God.* The same is true of the Hegelian and Marxist dialectics presumed to be driving history, as also the case of evolutionary-based progressive views.

By contrast, the God of the Bible allows for certain unpredictable and new things including the births of unique and interesting people even some who have changed history. Thus, only the God of the Bible who transcends time can know the inscrutable future! Though the God of the Bible has a plan for history and intervenes to change the course of history, his sovereignty is sufficient to give free will (self-caused acts) to those who he has made in his own image.

Absent the influence of the Bible, mankind reverts to some or another eternalist teaching, for example: a world of random chance as taught by the ancient Epicureans; a world of eternal cycles as taught by the Stoics; or mankind and the world as having always existed as taught by Aristotle. As Eliade points out, most cultures from ancient times until now have lived or reverted to what they assumed to be an essentially unchanging world.

Having rejected the Bible and its miracles, the Deist thinkers of the Enlightenment such as David Hume also forfeited a sense of history. As did the classical pagans, Hume wrote and studied history not to understand the present but to learn the eternal wisdom pertaining to the nature of mankind. The Deists idealized the classic (i.e. classed-based) civilizations of pagan Greece and Rome. Praising the classic pagan virtue of manliness, they regarded Christianity as weak and unmanly. Deism however was more congenial to what they saw as needed for the colonial dominance of the world's weaker peoples.

What is currently referred to as the 'Western way of thinking' is a combination of the historical way of thinking rooted in the Bible and the non-historical *eternalism* that is rooted in Greek philosophy. These contrasting ways of thinking have influenced Jewish, Christian, Islamic, as well as Western secular thought.

The historical way of thinking which comes from the Bible has very slowly shifted Christian theology and Western secular thought including science from the eternalist and historicist ways of thinking that come from pagan Greece! This is due to the ongoing influence of the God of the Bible over the course of Christian history. Thus, understanding the present world requires knowing its history, especially the history of ideas. Whether it be theology, philosophy, or scientific thought, very little can be understood apart from the history of ideas. Hence, the need to study scientific thinking from its historical roots as I have been doing throughout this book.

Historical understanding requires the preservation of history. While songs and poems can begin the preservation of history, books are far superior, though it is something that requires a literate people. Prior to the Reformation, literacy and learning in the Latin speaking West were in the hands of the clergy, the people of the Book. In Western Europe, a *clerk* (i.e. member of the clergy) was the same as one who could read and write.

Science, which began with Aristotle's philosophy, contrasts with the historical worldview of the Bible. During the early Christian centuries, science was proscribed as sorcery due to its associations with astrology, alchemy, and magic. Thus, the writings of Aristotle were not translated into Latin, the language through which Christianity and literacy spread to Western Europe. Rather, from the time of Augustine in the late fourth century, an otherworldly Neoplatonism with its ascetic despising of the material world dominated Christian theology. It was therefore among the Arabs of the Islamic East where arose the beginnings of science. It was not until the Crusades that Aristotle's writings, which Nestorian Christians had translated into Arabic, were translated into Latin and Aristotle's science as developed by the Arabs became known in the faculties of the arts in Western universities.

Following Rabbi Moses Maimonides (1138-1204), Thomas Aquinas (1225-1274) shifted Augustine's Neoplatonist theology from a Platonic to an Aristotelian basis, a worldview that could accommodate science. However, the Thomistic worldview as depicted in Dante's *Inferno* lacked a true sense of history. Rather, Aquinas medieval worldview reflects the static feudal order of pagan Rome. Thus, Renaissance artists depicted the people of the Bible in Renaissance dress. Aquinas was however challenged by the voluntarist tradition in late medieval philosophy which on account of the greatness of the God of the Bible, rejected the idea that the human mind can understand the world apart from studying the Creator's individual creatures (c.f. the discussion of *voluntarism* in Chapter 14). This led to the empirical and analytical traditions that led to the Reformation. Together with the mathematics stemming from the Renaissance revival of the teachings of Plato, this led to the Scientific Revolution which distinguishes modern from Aristotelian science.

THE PHILOLOGICAL ROOTS OF MODERN HISTORIOGRAPHY

It was the humanists of the Renaissance who recognized writings found in libraries as divided between *ancient, medieval,* and *modern*. By *ancient* they had in mind the classic civilizations of Greece and Rome. With these major divisions began our secular world's historical understanding. A shift to the historical worldview of the Bible began in earnest with the back-to-the-Bible Reformation. Seeking sources to argue who better reflected the teachings and practices of the early Christian churches, Protestant and Catholic historians together created a historical understanding of the Christian centuries.

The Scientific Revolution began as careful observations of the skies led to challenges of Aristotle's earth-centered cosmology and the paradigm shift known as the Copernican Revolution discussed above. Due to the Council of Trent endorsing Aquinas, Protestants were more willing to favor science that challenged Aristotle. Beginning with Joseph Scaliger (1540-1609) but more famously remembered in the chronicle of the nations of Archbishop James Ussher (1581-1656), this revolution gave rebirth to the chronology of nations that had been neglected since the time of the church fathers. Unlike the humanists whose 'ancient history' began with Greece and Rome, Scaliger gave attention to nations mentioned in the Bible such as Egypt, Persia, and Babylon. Due to being grounded in the Bible, this led to the Western world's uniquely historical thinking, even when the scientific history of man conflicts with that of the Bible as in the case of the present scientific narrative.

27

Calculating the Beginning of Egyptian civilization

Before the remains of the earliest humans were supposed as dating from the Ice Age, challenges to the antiquity of man as recorded in the Bible came from what some believed were more ancient records preserved by nations such as Egypt, India, China, or even from some civilizations of ancient America. But aside from pious Christians, nineteenth century scientists believed as did the ancient Greeks that Egypt was the oldest civilization. After deciding that northern Africa was under water from the melting waters of the Ice Age, geologists concluded that the Ice Age must have ended before civilization began in Egypt. But when did civilization begin in Egypt? Determining this would be a way of estimating the time since the end of the Ice Age.

Taking his cue from M. Rosiere's work to measure Egypt's oldest monuments following Napoleon's invasion of Egypt (c.f. Chapter 24), a scientific solution to the beginning of Egyptian civilization was worked out in the early eighteen fifties by Lyell's father-in-law, Leonard Horner, then President of the London Geological Society. From London, Horner directed drillings into the Nile River delta to determine the time that had passed since the erection of what he believed were Egypt's oldest monuments. The time would be calculated from the silt that had built up in the Nile delta covering these monuments. Based on the depth in which burnt bricks had been found beneath the Nile Delta, Horner concluded that it had been 13,371 years.[20]

How remarkably close this "scientific" measurement to the beginnings of Egyptian civilization which the Greek historian Herodotus

---

[20] An excellent history of Horner's work has recently been published. c. f. Meira Gold, 'Ancient History and the geological antiquity of man, 1847-1863,' *History of Science* 2019, Vol 57(2) 194-230

calculated by what he learned from an Egyptian priest.[21] The Egyptian priest told Herodotus that the Egyptians had enjoyed a continuous priesthood of 341 generations from the time of Egypt's beginnings as represented by the wooden statues that each high priest had carved for himself. Calculating 3 generations for each 100 years, Herodotus determined that Egyptian civilization had begun 11,340 years ago. Around 1850 when these soundings were conducted it had been more than 2,000 years since Herodotus learned this, bringing us too very near to what Horner calculated.

Herodotus's calculations based on Egyptian myth was however hardly more fanciful than Lyell's "scientific" measurements as the latter multiplied a 6-inch per century build up to a depth of 60 feet where Horner had found burnt brick(!), arriving at a neat 12,000 years for the beginning of Egyptian civilization. Lyell published this date in his influential 1863 book, *The Geological Evidences for the Antiquity of Man.*[22] This span of time continues to define the Holocene, the last and most recent geological era which follows the end of the Ice Age. Should anyone doubt that scientific dating is calibrated to accord with the accepted scientific narrative, know that more than 70 years since the beginning of radiocarbon dating, an era of 11,650 bp for the Holocene is recognized by the *International Commission on Stratigraphy!* Rather than robust and objective scientific dating as generally presumed, doubt not that the dating of ancient man is based on preserving the narrative!

---

[21] *Herodotus, Book Two*, p 186 of 1954 Penguin Classics translation

[22] Charles Lyell, T*he Geological Evidences for the Antiquity of Man* (2004 Dover edition), p. 30

28

Fossil Man and the Rise of Evolution

Though a Deist, or rather because he was Deist, Charles Lyell was much concerned to protect the dignity of man from a possible relationship with lower natural forms of life as might be the case with human evolution. Thus, Lyell celebrated his uniformitarian geology as having refuted evolution. To protect the dignity of man, Lyell wanted an absolute demarcation between geology and human history as could be determined from the history of man as given in the Bible. Lyell was thus a dualistic Idealist, a liberal Christian as deriving from the enlightened philosophy of Immanuel Kant (1724-1804). Following Kant, liberals separate moral truths which guide the behavior of mankind from the factual truths of science. Such a separation of the historical sciences from religion is however incompatible with Christianity where theology itself is rooted in historical events such as Creation and Jesus's resurrection from the dead. As was also the case of Kant, Lyell may have wanted to protect the dignity of white Europeans over the lesser colored races. In this regard he was no exception to the scientific minds of his time.

Keep in mind that from the time of the Enlightenment, especially Deists such as Lyell approved of slavery and like Agassiz and Darwin, he was deeply racist! Aside from Christians who regarded the Heathen as having descended from Noah, almost every educated person in the late nineteenth through the first half of the twentieth century was a racist! This was because racism was the reigning scientific anthropology! Then as now "progressive" Christians find ways of interpreting the Bible to accord with the latest science. Having little direct contact with colored peoples, as the case of Louis Agassiz before he came to America, racism appeared an obvious truth to educated Europeans.

Evolution was not invented by Darwin whose theory of evolution by natural selection was but the last of several theories of evolution. As Arthur Lovejoy pointed out, evolution came about through temporalizing the Enlightenment's Great Chain of Being. The close relationship between humans and primates had already been suggested by some versions of the Enlightenment's Great Chain of Being such as Carl Linnaeus (1707-1778) *System of Nature* which placed humans in the same class as primates. Evolutionary theories began with the Scottish Primitivists who taught man as developing in stages from woodlands to herding and farming, to a commercial economy as taught by Adam Smith. The Scottish Primitivists included the eccentric Judge James Burnett (1714-1799) who believed that humans were born with tails which midwifes discreetly remove before presenting babies to their mothers. Burnett was close friends with Darwin's grandfather Erasmus Darwin who put forth one of the earliest theories of evolution. One can understand why the young Darwin proposed man's ancestors as coming from African apes. Doubt not that the current scientific narrative of mankind's origins as being in Africa is due to Darwin's legacy which every kind of scientific dating will dutifully serve.

Fig 26 - Fossil Man (1838) conceived as dark-skinned, simian, and furry

Notwithstanding his aim to refute evolution, Lyell's *Principles of Geology* only inspired theories of evolution by acquainting the public with a long progression of life while not explaining the origin of species. The popular 1844 book *The Vestiges* of *the Natural History of Creation* created no small controversy in Great Britain. Alfred Loyd Tennyson's 1849 poem with the famous line 'nature red in tooth and claw' was inspired by Lyell's *Principles of Geology.* Herbert Spencer's teaching of *"Social Darwinism"* preceded Darwin's 1859 *Origin of Species*, but Darwin would embrace this idea in the subtitle of his book concerning the 'preservation of the favored races.' This explains why Lyell wished to deny the fact that remains of man were to be found among the fossils.

However, Lyell's chief motive in denying that the remains of man might be found in what he called the *Drift* was because it would serve as evidence that the Diluvium Flood was in truth the Noah's Flood of the Bible!

29

STRONGLY DENYING THE EVIDENCE OF MAN AMONG THE EXTINCT ANIMALS!

Should the deposits now being attributed to an Ice Age be those of Noah's Flood, we should expect to find evidence of man among these deposits. The failure to find such evidence was the reason that Adam Sedgwick and William Buckland eventually acknowledged that the Diluvium Flood could not be the deposits of the Flood of Noah described in the Bible.

Whether imaginative or real, there had long been reports of the remains of some who died in the Flood as the case of the 'man as witness to the deluge' proposed by the Swiss naturalist Johann Jacob Scheuchzer (1672-1733). Having pioneered the method of identifying fossils, Cuvier recognized Scheuchzer's specimen as the bones of a giant salamander. However, in 1771, Johann Esper found human bones in a cave with a giant bear and other extinct animals. Cuvier accepted Esper's judgment that the human remains were likely not from the era of the extinct animals. Cuvier also examined the celebrated skeleton encased in limestone that had been found in the Caribbean island of Guadalupe. Finding soft tissue in the skeleton, Cuvier turned from his initial open-mindedness to dogmatic insistence that there are no human fossils. Popularly known as the 'Pope of Bones' due to having founded the science of paleontology, Cuvier's judgment carried great weight in the international scientific community. Buckland had tied his scientific identity to the reputation of the great Cuvier.

The bones of the extinct animals found in his 'Last Revolution" which Cuvier identified with the historical Flood remembered by the nations were too recent to be petrified. Cuvier would have known this, especially in the case of the frozen mammoths found in Siberia which had flesh fresh enough to be eaten by wolves. However in its original sense the word *fossil* meant something that had been dug up. Thus,

Cuvier's dictum of there being 'no human fossils' was understood as there having been no humans living in Europe during the era of extinct animals. Cuvier and Buckland believed that at the time of the Diluvium Flood, mankind had not spread from the lands of the Bible

In 1823, Buckland himself found a human skeleton in the Diluvium deposits of the Paviland cave near Swansea in South Wales. This skeleton was coated with red ochre, probably having gotten such a coating due to red clay in the waters that once flooded the cave. The skeleton was from the same era as the extinct animals that Buckland famously championed as having perished in the Flood. As a devoted disciple of Cuvier, Buckland did not wish to challenge his mentor. Due to the shell necklace found on the remains, Buckland decided the skeleton belonged to a young lady from the Roman era whose grave had been dug into the Diluvium layer of the cave. The skeleton was thus celebrated as the *Red Lady of Paviland*. However in 1912, William Solace re-examined the cave where Buckland had found his Red Lady of Paviland, finding it to have been the bones of a young warrior from the era of extinct animals.

As early as the end of the seventeenth century, a Mr. Conyers had found a flint hand axe associated with the bones of a mammoth near Gray's Inn Lane. Then, in 1797 John Frere reported on hand axes at Hoxne near Dis in Suffolk which were buried 12 feet below the surface of the ground in undisturbed strata where were found the bones of extinct animals.

Fearing that evidence of man found in the Diluvium deposits might be used to argue that the Diluvium were the deposits of Noah's Flood, Charles Lyell became the strongest and most influential proponent in the scientific establishment holding to the claims of Cuvier and Buckland that no evidence of man existed in what Lyell was then calling *Drift*. By this time, most geologists were referring to the Diluvium as the *Pleistocene* deposits of a supposed Ice Age.

Denying that remains of man might be found in "the Drift" also suited Lyell's Deist conviction that everything associated with mankind including the ephemeral history of man put forth in the Bible was separate from the outdoor world of Nature that existed outside of history, a Nature that his Deist God had created for the benefit of man. Thus, finding evidence of man with extinct animals would also challenge Lyell's uniformitarian geology.

30

Then Recognizing Evidence of Man Found with Extinct Animals!

Among the evidence that Lyell had dismissed were chipped flints found in drift deposits in the Somme Valley near Abbeville, France that had been carefully excavated by Jacques Boucher de Perthes (1788-1868). Jacques de Perthes exhibited these in 1838, calling them *Diluvium* remains. Because of the cold reception that Lyell and the geologists gave his *Diluvium* remains, de Perthes published them again in 1848, this time calling them *antediluvian*. This slight change in terminology allowed that these human artifacts no longer need to have been from those who died in the Flood but possibly from far more ancient times, presuming that the Diluvium were the deposits of hundreds of thousands of years as Lyell had taught geologists to imagine them.

In 1858, an English geologist named Falconer was visiting the area and saw these chipped flints. After arriving back in England, he shared his observations with Sir John Prestwich and John Evans, recommending that the two of them visit the site in Abbeville, France. Prestwich and Evans did just that, spending much time at the site and confirming that without doubt there was evidence of man in the same drift deposits as were found extinct animals. Evans wrote to his sister, 'the flat axes and implements found among the beds of gravel were evidently deposited at the same time with them in fact the race of men who existed when the deluge, or whatever was the origin of these gravel beds took place.'

As his letter to his sister reveals, Evans understood that these might be evidence of men who perished in the Flood. But Lyell managed the narrative that would be used to understand these discoveries. On May 26, 1859, Prestwich read his paper to the Royal Society entitled *On the Occurrence of Flint Instruments with the Remains of Animals of Extinct*

Species in beds of a Late Geological Period at Amiens and Abbeville and in England at Hoxne. Prestwich's lecture was attended by Lyell, Huxley, and Faraday. Shortly afterwards at the British Association of Science, Lyell giving the Presidential Address announced that he was 'fully prepared to corroborate the conclusions ... recently laid before the Royal Society by Mr. Prestwich.'

This would be the same year that Lyell's disciple Charles Darwin would publish his theory of evolution. In his first book, Darwin avoided treating the evolution of man, but in 1863, Lyell published his influential *Geological Evidence of the Antiquity of Man,* which prepared the way for Darwin's 1871 *Descent of Man.*

What should have been heralded as the first acknowledgment of those who perished in Noah's Flood has instead long been heralded as the discovery of the great antiquity of man!

In the next chapter, I will explain how Darwin's neighbor John Lubbock distinguished these artifacts from similar ones found in Stone Age structures and thus date the former to a far earlier age. As I mentioned in Chapter 4, this dating was determined by geologists instead of archaeologists. These geologists ignored the fact that Diluvium also covered these Stone Age structures including the one in Fig 4 of Chapter 3 above. Otherwise, they would have recognized these as belonging to the same antediluvian world, none having great antiquity. The human remains and artifacts found in a Stone Age dwellings were essentially identical to those in the caves and in the Drift.

Instead of finding evidence of antediluvian man just as recorded in the Bible, Lyell would market the great antiquity of man. This confirmed doubts about the truth of the Bible that had begun when Lyell explained away the evidence of the Flood. Supposing that man was living on earth for hundreds of thousands of years before Adam challenged the most basic Christian theology. Rather than confirming the Bible in a major way as should have been the case, this opened the door for (1) the evolution of man and (2) consequently in the English speaking world, the German higher criticism of the Bible. In sum, this discovery shifted the modern world's principle intellectual authority from the Bible to Science -- from God to man! Ponder this important 1859 date.

In the public mind, this shift in intellectual authority from the Bible to evolutionary science would be associated with Lyell's disciple Charles Darwin. It was indeed Darwin's theory that was debated when this shift occurred. However, the controversy over Darwin's theory would mask the foundational shift in scientific thinking which had preceded Darwin's theory of evolution. Darwin himself understood how his theory depended on the near simultaneous "discovery" of the "great antiquity of man."

Due to discovering what appeared an extremely early period of man contemporary with extinct simians, all evolutionists needed for their theory was to suppose certain human remains (Neanderthals) as not being fully human and suppose certain simian remains as being man's immediate ancestors. As the case of all megafaunas accompanying antediluvian man, simian skulls such as gibbons are larger than living species. Underlying these factors was Lyell's clever way of explaining away the evidence of the Flood while showing Darwin how to proclaim "missing links" for supposed evolution in the fossil record. It was not so much Darwin but his mentor who is most responsible for changing the modern world's authority concerning origins from the Bible to Science. This was a world-changing shift. I doubt however that the pious Deist Charles Lyell was happy with Darwin's atheistic theory of evolution by natural selection.

31

DATING THE ANTIQUITY OF MAN

In 1859, when Lyell led the world in recognizing the remains of man in the Drift, something which he had so long denied, it been about thirty years since he had convinced most geologists that the one-year Diluvium Flood was an extended geological period which he called *Newer Pliocene.* By 1859, most were supposing this period as a long Ice Age referred to as the *Pleistocene*. But just how long was mankind supposed as having lived on earth and how would this be determined?

Keep in mind that what was now being viewed as an Ice Age consists of the buried deposits of the antediluvian world which date from only about six thousand years ago. However these antediluvian deposits may include the bones of animals and various organic remains that date before the creation of man.

Prestwich himself believed that what would be classified as Paleolithic remains may have been from those who perished in some Flood. He estimated the era of drift deposits as occurring over about a 20,000 year period.

In Chapter 27, I related how Lyell's method of estimating time was based on the depth of deposits assumed to have accumulated at a uniform rate. Thus, Lyell and his father-in-law Leonard Horner dated the beginnings of Egyptian civilization to about 12,000 years ago. This became what Lyell called the *Recent Period,* now called the *Holocene,* the time since the Ice Age. Lyell published this date in his influential 1863 book, *Geological Evidences of the Antiquity of Man,* a date that has hardly changed since 1863.

Based on the depth of the Somme deposits where was found evidence of man with extinct animals, Lyell supposed the era of these deposits as having been about 220,000 years! Here, Lyell was giving the first ever dating of the Ice Age. The *Pleistocene* (Ice Age) geological period was supposed as corresponding to what Lubbock would define as the *Paleolithic* period of man. Thus, a newly defined archaeological period was used to obtain a quantitative duration for a geological period called the *Pleistocene*, while the geological period was in turn supposed as dating the Paleolithic remains. If you are confused by the circular reasoning used for defining and dating the Ice Age, so have been geologists and archaeologists!

The deposits which Lyell supposed to have accumulated at a uniform rate are found at all manner of depth whether one recognizes them as having been deposited by the Flood or by ordinary geological processes. No matter his method, Lyell's pronouncements carried the ponderous weight of scientific authority. Thus, the duration of the antediluvian world which we can date from the Bible as being less than two thousand years, then expanded to what geologists were supposing as an Ice Age lasting for hundreds of thousands of years.

Sir Arthur Evans (1851-1941), son of the John Evans who had inspected the Somme deposits, used the same rate of deposition as Horner and Lyell had determined from the Nile delta to date the first Neolithic settlement in Greece as being between 12,000 and 10,000 BC. But why would the rate of the buildup of silt covering archaeological sites in Greece be the same as those covering sites in the Nile delta? Arthur Evans was adjusting his measurements to correspond to Lyell's dating. That is what archaeologists and scientists have been doing since.

The Pleistocene (the Ice Age) together with the Holocene (time since the Ice Age) form the *Quaternary*. In geological terms, the Quaternary corresponds to the era of mankind as described in the Bible. **The Pleistocene/Holocene division roughly corresponds geologically to the antediluvian/post-Flood eras of man.**

I will explain below how the Ice Age duration would expand to more than ten times Lyell's initial estimate, but not even Lyell would have suggested such a long period had not geologists failed to recognize that the "Ice Age" deposits covered not just Paleolithic remains but also Stone and Bronze Age dwellings! This was also Lyell's doing because more than anyone else except perhaps James Hutton, Lyell was determined to exclude the remains of man from the science of geology.

Note that the Paleolithic/Neolithic chronology was never based on strata but on an artificial typology which could be used to create chronological periods from artifacts in contemporary sites that might have been occupied for less than a generation!

32

An Impossibly Long Paleolithic Prehistory

The archaeological understanding of 'man in the drift,' also associated with Ice Age cavemen was put forth in 1865 by Darwin's neighbor and friend Sir John Lubbock (1834-1913). According to Trigger, Lubbock's *Prehistoric Times as Illustrated by Ancient Remains, and the Manners and Customs of Modern Savages* was almost certainly the most influential work dealing with archaeology published in the nineteenth century.[23] Lubbock's artificial Paleolithic/Neolithic chronology was established almost one hundred years before the advent of radiocarbon dating.

Lubbock's "archaeology "was based on anthropology, his study of living peoples. Influenced by Darwin's description of the Fuegians who Darwin had met on the southern tip of South America, Lubbock enjoyed documenting the nastiness of savages. Holding rigidly to the unilinear evolution of culture, Lubbock vehemently denied that degeneration of culture had ever taken place. He thus projected into scientific prehistory how he viewed living "savages" such as the natives of Australia.

Lubbock and his colleague Thomas Huxley who became Darwin's bulldog believed that native Australians were not fully human. Thus, Australian aborigines could be hunted and killed like animals to make room for progress. The British settlers who did this had Science on their side! Religion has rightly been seen as responsible for many evils, but Science has far more responsibility for great recent genocides including those of Hitler who was informed by this same science.

---

[23] Bruce G Trigger, *A History of Archaeological Thought* (1989, Cambridge University Press), p.114

Geologists from the time of William Buckland recognized the cave deposits as being the same as the Drift deposits. Thus, Lubbock determined that those who crafted the unpolished flint weapons found in the drift along the riverbanks had to be the same as the cavemen who were found with similar tools and extinct fauna. But instead of dating by supposed geological strata, Lubbock shifted to dating man according to his racist-inspired progressive development of artifacts.

Following geologists like James Hutton and Charles Lyell who had long excluded stone age houses covered by geological deposits from the study of geology, Lubbock determined the men found in the drift to have been much earlier and less advanced than those in the mounds who crafted similar rough-hewn tools but also owned a few polished tools. With his superficial typological criterion, he distinguished the *Paleolithic* (Old Stone Age) cave men from the *Neolithic* (New Stone Age) men buried in the mounds. However, Paleolithic and Neolithic Man lived during the same Early Bronze Age!

Believing in Social Darwinism, European archaeologists would reckon their white Europeans ancestors who originated in the Caucasus had killed and displaced the cavemen who occupied Europe during the Ice Age.[24] As did Lyell and most geologists, Lubbock failed to notice that the drift also covered the "Neolithic" mounds, megaliths, and kitchen middens which he so extensively discussed in *Prehistoric Times*. All these mounds were created by the same great Flood as had created the drift and cave deposits.

Instead of understanding these remains as the deposits of a great Flood lasting for one year, viewing these as the deposits of a long Ice Age meant that the humans using only un-polished tools maintained a primitive but unchanging technology for hundreds of thousands of years. That is an impossible accomplishment for tiny bands of hunter-gatherers who must be constantly on the move as local sources of food are soon depleted. Unlike animals, clothing for warmth and protection does not grow on human bodies. Human babies are also slow to mature thus a liability during their early years. How did these tiny bands fail to get wiped out by famine, freezing weather, flood, fire, drought, pestilence, disease, or killed by fierce animals or competitive bands seeking their provisions?

Paleolithic men could not have been very bright to have failed to learn how to domesticate the sheep and goats they are supposed to

---

[24] W. J. Sollas, *Ancient Hunters and their Modern Representatives* (1924, McMillian)

have followed or plant some of the grains they were gathering. It is all the more astonishing when we consider that Paleolithic man was highly skilled in the far more difficult tasks of making fires, clothes, leather sandals, blade-and-burin tools, traveling across the entire planet and learning to adapt to the most difficult environments as supposed to have existed during the Ice Age. In truth, these “Paleolithic” humans were either impoverished dwellers of caves on the outskirts of cities or else bands of hunters from the highly developed Early Bronze civilizations belonging to the antediluvian world.

Should these Paleolithic bands have lived over such a long period, they would have regularly returned to the most fertile camps, those near springs and rivers. Regardless their primitive technology and small numbers as could dwell at single sites, after hundreds of thousands of years, the refuse from their meals, shelters, campfires, and industries would have accumulated to a vastly greater depth than the shell middens that Lubbock assigned to Neolithic man. None of the Paleolithic sites show signs of having been occupied for so long.

One of Lyell's disciples, Edouard Lartet (1801-71), learned to date these Paleolithic findings according to what he supposed as the flourishing of various now extinct animals whose remains were discovered in caves throughout Europe. (1) According to Lartet, the earliest men found in the Drift belonged to the Age of the Cave Bear. (2) Then came the Age of the Mammoth and Wooly Rhinoceros. (3) Next came the Age of the Reindeer. (4) Last, came the Age of the Auroch (wild cattle) and Bison. An even earlier Hippopotamus Age was soon added. Following the scientific methodology of his mentor Lyell, Lartet pulled geological eras like rabbits from his zoological hat. Though supposed arranged according to zoological-based geological periods, Lartet's periods were based on what he imagined as a progressive development of artifacts.

Gabriel de Mortillet (1821-98) continued Lartet's work, though replacing the names of Lartet's zoological eras with the now familiar archaeological ones. The Age of Hippos became the Chellean. The Cave Bear and Mammoth Age became the Mousterian. The Reindeer Age became the Magdalenian. Due to his belief in unilinear cultural evolution, Mortillet searched for signs of whether what appeared as ordinary rocks (eoliths) might instead indicate mankind's first use of tools.

However in 1880, a little girl found highly skilled drawings of mammoths and various extinct animals on the walls of caves in southern France. Such skillful and beautiful art on the part of primitive cavemen hardly suited Mortillet's dogma of unilinear evolution. Thus, the leading Paleolithic archaeologist refused to recognize these drawings as authentic until the end of the nineteenth century when at last he agreed to assign them to the most recent era of the cave men. This was not however the only discovery of the late nineteenth century that was challenging the dogma of unilinear evolution.

Unlike Mortillet, a materialist and strict unilinear evolutionist whose work was based on museum collections, the Catholic priest Henri Breuli (1877-1861) was willing to recognize influences from the Ancient Near East. Breuli introduced the *Lower, Middle,* and *Upper* classification of Paleolithic remains. But regardless his "Lower," "Middle," and "Upper" classification, strata was no more used for dating the Paleolithic than for the following Neolithic period. There is no valid reason for supposing that all these eras were other than concurrent. The larger skulls of the mega-fauna apes and gibbons gave plenty of candidates for a supposed "Lower" Paleolithic. The cavemen were of course found in caves, some using them as temporary hunting shelters (thus modern humans) and the more impoverished, having developed rickets due to these caves being their permanent homes (thus, Neanderthals).

Recent DNA tests show that modern humans have Neanderthal DNA. It may be scientific myth that humans can mate with non-humans, not to speak of it being a vile racist assumption. Rather, these genetic tests reveal that Neanderthals were fully human because they mated with humans. Every human is made in God's image. Racism is no less wrong when applied to people who are now dead!

Currently, the world's oldest humans are supposed as living in Africa about 300,000 years ago. Anthropologist can find whatever date they wish for the earliest humans. But supposing humans descend from African simians is due to dutiful worship of Charles Darwin whose evolutionary theory serves as the gospel of naturalistic science. Neanderthal DNA however is missing from modern African populations.

33

Discovering Multiple Ice Ages and their Supposed Causes

Having failed to recognize the Diluvium as being the deposits of the Flood, in which case it is unsurprising that geologists should find layers of loess and trees in the Pleistocene deposits of Europe and America, these organic deposits suggested instead that there were periods of warming during the Ice Age. This led to the conclusion that there had been several Ice Ages.

Of major concern for those teaching an Ice Age was finding a cause for this great change in the earth's climate. One possibility was Lyell's theory that the cooling of continents was the result of their rise in elevation. This is why Lyell eventually accepted Agassiz's Ice Age teaching which he had long opposed due to it conflicting with his uniformitarian principles of geology.

Isaac Newton had explained how the earth's seasons depend more on changes in the earth's tilt than upon the distance between the earth and sun as the earth moves around the sun in an elliptical path. Thus, James Croll(1821-1890) of Scotland suggested that Ice Ages were caused by periodic changes in the earth's tilt. So impressed were geologists with Croll's theory, they abandoned Lyell's theory. Lyell's reputation suffered an eclipse until he revived it with his proclamation of the "antiquity of man."

Croll's theory however needed to be tested by measuring it against geological evidence of the beginnings, endings, and durations of the various Ice Ages. In 1909, two European geologists, Albrecht Penck (1858-1945) and Eduard Bruckner (1862-1927), believed they had found ways of identifying 4 Ice Ages which were respectively named for four European rivers (Gunz, Mindel, Riss, and Wurm) whose raised banks were supposed caused by the melting waters of these 4 Ice Ages. They calculated the duration of these Ice Ages based on the current erosion of gravels from these rivers. Penck and Bruckner calculated the total duration of the Ice Age as being about 650,000 years.

Though using different methods than Penck and Bruckner, American geologists also believed that they could identify four separate Ice Ages from sequential drifts which were from oldest to youngest were the Nebraskan, the Kansan, the Illinoian, to the Wisconsin, which was the most recent.

According to Croll's theory, the last Ice Age should have ended 80,000 years ago. I have already mentioned how the American geologists had found a way of calculating the time since the end of the Ice Age according to the erosion of Niagara Falls as having been no more than about 10,000 years. As put forth in the influential geological textbook by Yale geologist James Dana (1813-1895), American geologists strongly opposed Croll's theory.

A more refined and complex astronomical calculations of changes in the earth's tilt was proposed in 1924 by the Serbian engineer Milutin Milankovitch (1879-1958). Milankovitch pointed to 3 recent Ice Ages dating to 25,000, 27,000, and 115,000 years ago, then to 6 glacial minimums occurring in pairs over the last 650,000 years. Obviously, Milankovitch was aiming to fit his astronomical-based periods to Penck and Bruckner's geological periods! But finding geological methods to confirm these Ice Ages if even possible would be incredibly complex.

With the advent of radiocarbon dating of the drift deposits by Yale geologist, Richard Flint (1902-1976), geological opinion turned against Milankovitch's calculations, though having nothing better to define the Ice Ages, Milankovitch's calculations continue to inform Ice Age periods. In Chapter 60, I will cover Ice Age dating which is based on supposed reversals in the earth's magnetic field, which would be used to extend the Ice Ages to 2.4 million years.

34

PROBLEMS WITH THE MANY FORMS OF GEOLOGICAL AND SCIENTIFIC DATING

In the late nineteenth and early twentieth century, Scandinavian geologists developed a system of dating worldwide climate based on pollen found in successive layers of peat. The key pollen in this system is that of the Arctic Dryas flower. The *Blytt-Sernander* climate periods shown in Table II below have been used to date the late Pleistocene and early Holocene periods from before the advent of radiocarbon dating. By comparing the dates of Table II with Table I of Prehistoric Chronology discussed in Chapter 5, one can see how prehistoric archaeology continues to be based on the Blytt-Sernander system.

The Blytt-Sernander system presumes that layers accumulated at approximately the same rate prior to and following the end of the Ice Age. However, the presence of the Dryas pollen is due to the cooler climate prior to the Flood. The layers of peat lower and earlier than the so-called Atlantic layer were deposited about the same time by the Flood which lasted only one year. For the reasons I explained in Chapter 4, radiocarbon

Table II

Blytt-Sernander Climate Periods

Pleistocene	Older Dryas stadial	14,000 -13,600 BP
Pleistocene	Allerod interstadial	13,600 - 12,900 BP
Pleistocene	Younger Dryas stadial	12,900 - 11,640 BP
Holocene	Preboreal	11,640 -11,500
Holocene	Boreal (cool, dry, rising tempertyres	11,500 - 8,900 BP
Holocene	Atlantic (warm, moist, max' temp')	8900 - 5700 BP
Holocene	Subboreal	5,700 - 2,600 BP
Holocene	Subatlantic	2,600BP - Present

BP = years Before Present

dating hardly changed this pre-established system of prehistoric dating.

The Blytt-Sernander system was used to support the idea that a period of warming immediately preceded the last Ice Age when the lowered sea level created a land bridge between Asia and North America. These dates were supposed confirmed and refined by the CLIMAP (Long range Investigation, Mapping, and Prediction) project of the seventies and eighties based on measurements of oxygen isotopes in sea floor deposits. Unsurprisingly, controversies continue to beset the CLIMAP project, which has become embroiled with that of climate change. The chief advertised consequence of supposed climate change does indeed rest on sea level changes caused by the supposed Ice Age. This however led to the currently understood rapid stadial/interstadial modifications of Ice Age periods supposed based on Milankovitch's complex calculations of Ice Age cycles.

Notice how the scientific dating game is played. Modifications of the scientific narrative of the transition from the last Ice Age are built on confirming the current narrative of the past. In this case, we see how the current dating was built on the supposed necessity of a land bridge between Asia and America that dates from the seventeenth century together with Charles Lyell's dating of the Holocene from the Greek historian Herodotus' myth-based estimate of the beginning of Egyptian civilization. The old narratives are confirmed by accepted patterns in the tea leaves of complex data and scientific argot.

Consider ice-core dating from sections of ice taken from bores in ice-capped Greenland and Antarctica going back to what we are told is about 800,000 years. Why don't they go back to what we are told is the beginning of the Ice Age about 2.4 million years ago? Whatever the period, would not the Flood have either melted or else added depth to the permanent ice? Was there even an ice cap prior to the Flood? This is a worthy project to research. Some claim to have found evidence of ancient melting. However annual deposits of snow are impossible to detect at depths where the ice load fuses the annual layers. Nor can we assume that such layers as might be counted are in fact annual layers.

In Chapter 4, I explained how radiocarbon is used for dating archaeological sites. But I did not there cover dating based on tree rings, ice cores, and lake varves which together with sediment and ocean floor bores have been used to measure ancient climate and "disprove" a worldwide Flood. The problem with deep ice-cores is that the weight of ice fuses annual layers. Thus, we are never shown what are supposed annual counts of ice layers!

We are told that some of these dating methods have been cross-checked by radiocarbon dating and correlated by deposit bands of historical volcanic eruptions as in the case of the Lake Suigetsu Project in Japan. Should this correlation of dating be as straightforward as advertised we should be very impressed. As I write this, I am hearing of an earthquake in this same region of earthquake and tsunami prone Japan. Can you believe that the bottom of this lake has never obtained deposits of new sediments from tidal waves and has never experienced erosion from the same over the last 50,000 years? One should never confuse neat illustrations with sound methods of dating.

Counts of the annual growth rings of trees reveal deviations in radiocarbon dating. These are used to calibrate radiocarbon dates. Based on the tree ring calibrations, it seems that in the centuries after the Flood, radiocarbon provides dates are too young. I suspect this is somehow related to the change in atmospheric radiocarbon due to the effects of the Flood. Whatever this effect, calibrated radiocarbon dates should be supposed reliable from the time of the Flood. But older dates may be suspect because tree ring sequences from different trees are themselves based on radiocarbon dating, a process of circular reasoning! Wiggle-path matching for correlating these sequences is, to say the least, hardly straightforward, most especially when done by computer!

There are other kinds of scientific dating including thermoluminescence used for measure ancient heating events. Whatever the method, all forms of dating require some means of calibration, supposed provided by tree rings or radiocarbon, but which themselves need calibrating for years earlier than the Flood. Reliable dating earlier than the historical period or since what is now understood as the beginning of an Ice Age does not currently exist.

No one should suppose that my criticism of these forms of dating are to defend a young earth! I do hope to see a more reliable dating methods for the recent history of the earth. The recent discovery of Noah's Ark, combined with the understanding of Noah's Flood explained here, may result in more accurate scientific dating, at least for the era of mankind. What is most needed in the practice of this kind of science is hard-nosed skepticism. However, there has never been a more domesticated group of scientists than those devoted to dating things from the Ice Age. But otherwise, the science of dating things from the Ice Age would implode.

35

The Roots of Racism and its Impact on the Prehistory of Man

As I mentioned in Chapter 4, a racist anthropology informs the prehistoric chronology of the scientific narrative of early man. Being the same as the Enlightenment's *scientific* anthropology, racism predates belief in the antiquity of man and thus served as the view of the scientific establishment when developed the scientific prehistory. In contrast to this racist scientific anthropology, the core teachings of Judaism, Christianity, and Islam eliminate the distinction of races within these related faiths, all of which recognize that all humans descend from Adam and that all nations descend from Noah's three sons only less than five thousand years ago. Due to not descending from Noah or his wife, the three wives of Noah's sons explain the differences in color of people from Asia, Europe, and Africa. All nations are of the single family who once lived aboard the Ark. One can thus understand how the deeply racist teachings of the nineteenth-century *American school of anthropology* lamented the fact that the Bible was holding back the "progress" of science.

Though we tire of hearing racism used to explain every real or imagined injustice, it is necessary that I address the roots of true racism and how this led to the prehistory of man that is so contrary to the origins of man put forth in the Bible. Knowing the true roots of racism is also helpful for healing racism. Here, I can only summarize how teachings coming from philology, nationalism, biology, ideology, and even theology lie behind racism. While these roots are important for understanding subjects as varied as eschatology and economic ideologies, my aim here is to explain how these racist teachings gave birth to the scientific narrative of early man.

While we have become accustomed to thinking of race in terms of the color of one's skin, the earliest racism was based on language. The origins of racism understood as those speaking one language regarded as superior to those speaking other languages is to be found among the pagan Greeks who explained why barbarians are innately suited to slavery.[25] According to Plato and Aristotle, it was because of the inferior geography of the barbarian's native lands. Aside from vaguely remembering the Flood and their ancestor Iapetus (Japheth, son of Noah), the pagan Greeks failed to remember that all mankind had descended from Noah's family. Though some Greeks were interested in what might be the oldest language, Aristotle taught that the nations had always been in their homelands. Pagan Romans who inherited this same culture were no less racist.

Though coming from many nations, Christians made no distinction between nationalities, classes, or former religion. Rather, as the Apostle Paul wrote concerning the identify of those in Christ.

Here there is no Gentile or Jew, circumcised or uncircumcised, barbarian, Scythian, slave or free, but Christ is all, and is in all. [Colossians 3:11]

Christianity effectively eliminated the racism of the pagan Greeks and Romans, even though the earliest church fathers also spoke the Greek language. Ethiopians were among the first Christians.[26] Most of the leading early church fathers including Clement of Alexandria, Tertullian, Origen, Cyprian, and Augustine were African. The Jewish/Gentile worldview of the Bible simply shifted to Christian/Heathen, in both cases, a religious rather than racist identity. As one may note from King James Bibles, *Heathen* has the same meaning as *Gentile*. Even so, under the Old Covenant of Moses anyone could be converted from Gentile to Jew through circumcision and under the New Covenant of Jesus, from pagan to Christian through baptism.

Regardless the clear teachings of the Scriptures, Christians can and have been racist and have defended their racism by re-interpreting the Bible so as to conform to their racism.[27] Of course, Christians do the same for many unbiblical beliefs including evolution and what are now called "progressive" beliefs. In doing so, they shamefully violate the clear teachings of the Scriptures.

---

[25] c.f. Benjamin Isaac, *The Invention of Racism in Classical Antiquity* (2013, Princeton University Press)
[26] Acts 8:27
[27] c.f. Colin Kidd, *The Forging of the Races: Race and Scripture in the Protestant Atlantic World. 1600-2000* (2006, Cambridge University Press)

It should however not be surprising that the first moderns to teach racism, even polygenetic views of mankind, were early skeptics such as Paracelsus (1493-1541) and Bruno (1548-1600) who also held to other unbiblical views. Accordingly, these impious men were widely condemned by Christians.

There are two fundamental sources of the re-invention of racism within the large body of Christendom. Not surprisingly, both were causes by Europeans distancing themselves from their biblical and Christian roots. One of these sources was the admiration of the racist classical Greeks and Romans by the men of the Enlightenment. This however came about due to Christians beginning to view the Old Testament as a Jewish rather than a Christian book. So did the ancient Gnostics, though this was something that neither the Apostles, the church fathers, Catholics, nor even most Protestants did until the twentieth century. Medieval commentaries viewed the Old Testament as Christian history as did America's Christian fathers who saw America as God's New Israel. This pleased Jews who enjoyed religious freedom in this New Israel. However by the beginning of the present century, the traditional Christian view would be lamented as "replacement theology."

TURNING THE JEWISH RELIGION INTO A SUPPOSED GENEALOGICAL RACE

This change came about due to certain Reformed Christians coming to view Jews as a race, though in biblical times as also in Rabbinic Judaism, Judaism is correctly understood as being a religion to which people of any race may convert. Viewing Jews as a race turns Christians into Gentiles, which is equivalent to regarding ourselves as Heathen! In this case, the identity of neither Christians nor Jews are in Christ, the Creator and Heavenly King of Israel which Christians believe to be the only way to God the Father. If the Old Testament is not a Christian book, Christ was the pretender that Rabbinic Jews claim him to be.

However, the Son of Man's primary identity was always in Heaven even when incarnated as a Jew. Not only did God the Father whom no one had ever seen use his Son to create the world, he also used Christ to call Abraham and deliver Israel from Egypt. As the Heavenly King of the Jews, Christ is the author of the Old Testament faith.[28] Christ however was only King of the Jews until his resurrection from the dead, at such time the Father granted to him all authority in Heaven and earth.

---

[28] c.f. 1 Peter 1:10-12

Unfortunately, racial identity can be read into biblical faith. The introduction of racism into Christianity came just after the Reformation due to identifying the Jewish religion as a race speaking the biblical Hebrew language. Those who did this did not seem to know that until about the tenth century, rabbinic Jews did not speak Hebrew. Even from the ninth century, Jews used the Hebrew language only for synagogue worship and study. Rather, across the ancient world Jews and Christians and later Jews and Muslims have spoken the same languages. This remained the case until the recent adoption of a new form of the Hebrew language by the modern state of Israel. From the Second Temple era (the time of Christ), most Jews understood Aramaic or Greek, neither being the Hebrew language in which most of the Old Testament was written. The early Christians used the Greek Septuagint which was translated by Jews and for centuries used for worship throughout the Roman Empire by Jews and Christians alike. I will explain below the introduction of Hebrew Torah scrolls for synagogue worship beginning around the ninth century.

Old Covenant Judaism was not based on race but on circumcision, a powerful sign that God's chosen were not his people simply by natural birth! Rather, Israel was distinguished from the Gentiles as being a peculiar, that is a religious nation rather than being a gentilic nation having a common natural ancestor. Beginning with Abraham, Gentiles were adopted as God's Chosen People through the covenant of circumcision. Thus, every male child had to be circumcised or else cutoff from Israel. Foreigners who lived among the Israelites were also circumcised. The Godly Uriah, husband of Bathsheba who was the mother of King Solomon, descended from the Hittites. This explains how the faith of the Jews could survive a dispersion, even obtaining many Gentile converts as during the time of Queen Esther.[29] At the time of Jesus, proselytes were famous rabbis. Thus, the Apostle Paul references those of the Jewish religion as the *circumcision,* whether or not they recognized Jesus as their Messiah. Paul distinguished the children of the faith of Abraham from those such as the Ishmaelites and Edomites who descend from Abraham by natural birth.

This is not to deny that genealogy was essential to the line of David as well as for priests and Levites. Priests were forbidden to marry outside the tribe of Levi so as to prevent intermarriage with powerful families, thus impacting the decisions of the ruling priests. This ancestry however was through the father rather than the mother as is

---

[29] c.f. Esther 8:17

the case of today's Rabbinic Judaism. David's lineage included several Gentile ancestors who had become Israelites by converting to the God of Israel. Even in the case of Levirate marriage for raising a family for one's deceased brother, legal genealogy differed from biological paternity.

Throughout the Bible, God's concern regarding intermarriage was one of religion rather than ancestry. Thus, an Israelite man sinned by taking the daughter of a foreign god as did King Solomon. On the other hand, as was the case of Rahab the Harlot and Ruth the Moabitess, a Gentile woman could become part of Israel through converting to the God of Israel. The failure of Christians to understand and teach this has resulted in the Old Testament being viewed as racist. Abraham was chosen to become the father of God's priestly Kingdom to bless all the nations! This is because the God of the Bible is also the Creator and the God of all mankind!

Though the Gnostics rejected the Old Testament, so much did Christians view themselves as God's Israel that it was Christians and only Christians who cherished and preserved the Jewish intertestamental as those of the Maccabees, also known as the Apocrypha. Michael Stone, the famous Jewish scholar of the Apocrypha, tells us that the oldest copies of these intertestamental Jewish books are in Armenia, the oldest Christian nation. After the Emperor Constantine made Christianity the official religion, Jewish proselytizing became difficult in the Roman Empire though not in the Caucasus and among the peoples north of the Black Sea and the Roman Empire which would become the cradle for Ashkenazi Jews. Thus, did the vast majority of Jews became the most "Caucasian" of all the world's people. Due to living in Islamic lands, Sephardic Jews made few converts aside from their slaves.

While Jews and Christians in the East and outside the Roman Empire continued to use the Aramaic (Syrian) language spoken by Jesus, the Hebrew text of the Old Testament was preserved only in Babylon. Though not recognized as Jews by the rabbis, the Karites of Babylon who reject the Talmud preserved the Hebrew text of the Old Testament, adding vowel markings for pronunciation. Following the Islamic invasions, Hebrew Torah scrolls spread to Rabbinic Jews who began adopting Hebrew for use in synagogue worship. Thus, the oldest copies of the Masoretic texts outside the Dead Sea Scrolls date from this period.

Even from Israel's earliest days, speaking the Hebrew language, a variant of Western Canaanite, was never the same as being an Israelite. Rather, the children of the patriarchs learned what the Bible calls 'the language of Canaan' due to having migrated to the land of Canaan.[30] The roots of early Israelite names, point to the original Hebrew language as being similar to the "non-Semitic" Hurrian language of Syria, from which land the patriarchs came. The Hyksos shepherd kings of Egypt who Josephus identifies as Israelites had Hurrian names. Ur of the Chaldees was not Ur of Shinar as generally presumed. Though beyond the Euphrates River, the home of the Chaldees was ancient Armenia where was once spoken the Hurrian language.

Wishing to translate the Bible into the vernacular languages from the original Hebrew, the Reformers went to the Jewish rabbis for Hebrew texts as well as to learn the Hebrew language. They learned from the rabbis who taught incorrectly that not only had Jews always spoken Hebrew, but Hebrew was also mankind's original language. Though mistaken about this, the rabbis were correct in seeing the Gentile languages as beginning with the confusion of languages at the scattering from Babel. Biblical "Hebrew" is thus a formerly Gentile language as is the Greek language that was used for writing the New Testament.

The earliest Christians were Jews who believed that Jesus of Nazareth was Israel's Heavenly Lord for whom John the Baptized prepared the way. Jews supposed that some if not all converts from the Gentiles were of the 12 tribes who God had scattered to all nations. Whether from the 12 tribes, all converts had to be grafted into the olive tree of Israel to serve as a light for the Gentiles. When referring to the diaspora, as was also the custom of Jews, the New Testament writers reference Gentile converts as among the 12 tribes of Israel.[31] Christ's Church is built on the foundation of the Apostles and Prophets, the former ruling over the 12 tribes of Israel.

Thus, from the beginning, Christians recognized the Old Testament as their own Scriptures, the only Scriptures quoted in the New Testament. Christians understood Jesus as the New Covenant fulfillment of the Old Testament prophecies if in a spiritual rather than fleshly sense, as many then and now suppose. Thus, the New Testament teaches the Promise Land as being Heaven[32] and God's

[30] Isaiah 19:18
[31] 1 Peter 1:1
[32] c.f. Hebrews 11:10

Temple as the body of those baptized in the Spirit.[33] This is why the spiritual hymn references the passing of believers as 'crossing Jordan.' The New Jerusalem is a reference to the Heavenly Church, who Israel's Savior has gathered for salvation. One enters the New Jerusalem (Christ's Heavenly Church) through gates having the names of the 12 tribes.[34] Just as the Old Testament Scriptures promised, Christ brought scattered Israel to the Promise Land which however is a Heavenly rather than earthly place until Heaven itself descends to earth.

Thus, Christianity and Rabbinic Judaism are of the same age, both rooted in Old Testament Judaism with one accepting and the other rejecting Jesus as the promised Christ. The majority of the Old Covenant Jews rejected Jesus as the Messiah. In their second century war with Rome, the leading Jewish Rabbi Akiva proclaimed Simon Bar Kochba as the Christ. After that disaster, messianic prophecy became unpopular among the Jews.

Perhaps inspired by the example of Mohammed, the rabbis returned to looking for a future Jewish Messiah as a military leader like Mohammed or King David who world restore the Jews to Palestine using an earthly sword. Far from holding "Judeo-Christian" values, rabbis do not subscribe to Jesus's Beatitudes, especially his teaching about turning the other cheek. However, the rabbis correctly recognized that according to the Scriptures only the Messiah could or should return Jews to Palestine.

THE ROOTS OF TODAY'S CHRISTIAN AND JEWISH ZIONISM

In turning to the rabbis to learn the Hebrew Scriptures, humanist scholars also learned their this-world historical interpretation of Old Testament prophecies. French rabbis such as Rashi (1040-1105) and Abraham ben Ezra (1092-1167) rejected Jesus as having fulfilled the OT prophecies. A this-world interpretation initially appealed to some Reformers because, wishing to avoid the excessive allegorizing of the Catholics, the Reformers aimed for literal interpretation of the Scriptures. However, a fleshly or this-world historical interpretation of OT prophecy conflicts with a literal reading of the New Testament. This is because Jesus and the Apostles taught a spiritual and Heavenly

---

[33] Ephesians 2:21

[34] Revelation 21:12

Kingdom as literally fulfilling the Old Testament prophecies which the prophets themselves had longed to understand.[35]

A spiritual understanding is the same as reading the Scriptures in the light of Christ.[36] But scholars and theologians who do not read the Old Testament prophecies in the light of Christ regard a literal reading as a fleshly and this-world fulfilment as did these rabbis who oppose Jesus. The Reformers thus understood that a this-world reading would require rejecting the New Testament interpretation of the Old Testament prophecies in favor of the interpretation taught by Jews who regard Jesus as a deceiver. Thus Martin Luther and John Calvin retained the traditional Christian prophetic interpretation in which Old Testament history serves as types and shadows for the Spiritual and Heavenly Kingdom of God.[37]

Even so, in the New Testament, the words *Israel* and *Jew* are found in the Bible far more than are *Christian* and *Church,* so much so that if the body of Christ is not God's Israel, Christians hardly have words of God written to and for Christ's Church. This however does not mean that the words *Jew* and *Israel* which continue to apply to religion and a nation going by those names today do not remind even unbelievers of the people of the Bible. Thus, did some Reformers including Calvin's successor Theodore Beza adopted a new interpretation of Romans 11. Traditionally, the passage '*in this way,* all Israel will be saved' had been read as coming about due to the full number of Gentiles coming into Israel by being grafted into the olive tree referenced in the context of this passage. According to the Apostle Paul, Jesus was the Deliverer who turned ungodliness from Jacob.[38]

However, Beza and the notes of the Geneva Bible interpreted the passage 'and then (that is, after the full number of Gentiles have come in) all Israel will be saved' as being a reference to a future conversion of the Jews, no longer understood as a religion that had rejected Jesus but now seen as a Hebrew-speaking race! Following this, it became common to refer to unbelieving Jews as "Hebrews" and as the unique children of Abraham. Thus did these theologians introduce racism into Christianity, something that would bloom during the Enlightenment. Not only did this turn Rabbinic Jews into a race (ethnos), it turned

---

35 c.f. Matthew 13:7, 1Peter 1:10-12

36 c.f. 1 Corinthians 2: 10-16

37 c.f. Alister McGrath, *The Intellectual Origins of the Reformation* (1987, Blackwell), chapter 6

38 Romans 11: 25-27

Christians into Heathen (Gentiles)! Instead of eliminating racism as had been the case from the birth of Christianity, this way of thinking would become the root source of racism which in the case of Jews could not be eliminated even by conversion to Christ! According to such thinking (that is, viewing the Jewish religion as a race) Christianity would no longer serve to eliminate racism but rather serve to promote racism! Thus, Christian/Heathen would be replaced by various kinds of racism all of it stemming from understanding Judaism as a race instead of being a religion.

This gave birth to what is known today as *Christian Zionism* and today's term, *Messianic Jew*. Whether there might someday be an earthly Zion in the lands of the Bible as in the case of modern State of Israel is a separate matter but note that Christian Zionism is centuries older than John Nelson Darby (1800-1882) to whom most theologians and scholars attribute this teaching. In truth, all Christians are Messianic Jews while God's holy hill of Zion is neither in Jerusalem nor Samaria, but are those who worship God in Spirit and truth. However this change to interpreting the Old Testament prophecies in the manner of the rabbis was taught by a few British theologians at the beginning of the seventeenth century to the great consternation of King James I, angered by those who taught that the English king should ever subject himself to the Jews.

This interpretation however also resulted in the Puritans rejecting the amillennial view that Catholic Church had held since the time of Augustine for a variety of pre-millennial interpretations. The millennium refers to the thousand-year rule of the saints mentioned in 20th chapter of the book of Revelation. However, keep in mind that until the rise of Zionism among unconverted Jews near the beginning of the twentieth century, this was seen as an event in Christian rather than Jewish history. Moreover, a premillennial view does not entail such an interpretation of Romans 11. The early church fathers were premillennialists as have been many Christians since who understand all prophecies as being fulfilled in Christ from his First to his Second Coming.

THE FORGOTTEN ISAAC LAPEYRERE, HERO OF AMERICA'S RACISTS

How earthly Zionism reintroduces racism was immediately revealed in the Zionist teachings of the Calvinist Isaac LaPeyrere (1596-1676). Coming from a family of converted Jews, LaPeyrere sought help from the Calvinist Prince of Conde, then after his conversion to Catholicism from the Pope, to restore the Jews to Palestine. Though ignored today, LaPeyrere, one of history's first Zionist, was infamous for teaching that all Gentiles came from men who lived before Adam. He justified this by pointing to stone tools used by the earliest inhabitants of France while the sons of Adam had tools of bronze and iron.[39] If Gentiles came from men who lived before Adam, their antiquity could be far older than Adam.

The idea that there were men before Adam from whom Cain obtained his wife wasn't original with LaPeyrere. It is at least as old as Julian the Apostate (331-363) who believed the same. What was new and radical about LaPeyrere was his teaching that the Bible concerned only Jewish history, that Adam himself was Jewish, and that Noah's Flood was limited to the land of the Jews. According to LaPeyrere, this explained why the Egyptians, Chinese, Hindus, and even nations in America had histories dating long before Adam. So beloved was LaPeyrere by defenders of racism, that Southern racists and Cornell University founder, Andrew Dickinson White who wrote *A History of the Warfare of Science with Theology in Christianity* (1896) regarded LaPeyrere as a martyr to the scientific truth of racism. An early defender of Darwin, White understood how profoundly influential were the teachings of LaPeyrere in shifting the understanding of the origins of man from a biblical to a scientific basis. White lamented the fact that the biblical teachings of nations as coming from Noah's family was holding back science! Unfortunately, it did not succeed in doing so.

As to why the Scriptures did not teach the same, LaPeyerre claimed it was because they had been corrupted, a view soon to be accepted by some of LaPeyrre's contemporaries including Thomas Hobbes, Richard Simon, and Spinoza. Of course, long before this, the Muslims learned from the Jews that Christians had "corrupted" the text of the Bible. Explaining the Bible by means of composition criticism was soon to be embraced by Protestants who as already the case with the Catholics were more than happy to base authority on the teachings of the established churches instead of the literal words of the Bible as was the practice of despised separatists and dissenters who would find religious liberty in the wilderness of America. Combined with the new understanding that Jews were a race defined by the Hebrew language, LaPeyerre's teaching resulted in what scholars understand as the German invention of racism.[40] However these newly enlightened

---

[39] c.f. Genesis 4:22

[40] c.f. *The German Invention of Race (2006)*, edited by Sara Eigen and Mark Larrimore

Germans were preceded by the Scottish Primitivists of Scotland's Enlightenment and by the French Enlightenment embraced by the Prussian King Frederick II (1740-1786). Thus, the savage American natives and black Africans were no longer children of Adam. They might not be fully human.

VIEWING THE OLD TESTAMENT AS A JEWISH RATHER THAN CHRISTIAN BOOK

Viewing Jews as a race rather than a religion would profoundly impact how many scholars including many Christians would come to understand the Old Testament no longer as Christian but as a book about those who became unbelieving Jews! Most Christians are not aware that Jews regard the Talmud which they refer to as the *Oral Torah* as being more sacred than the Old Testament, even the written Torah of Moses. After the destruction of the Second Temple, it was impossible to worship according to the Covenant of Moses. Even so, Christians have been taught to look to rabbinic teachings for understanding the Old Testament. How much light should we expect from those who view the light of Jesus as darkness? Rather, the Holy Spirit which comes from believing in Jesus is necessary for understanding the Old Testament.

Christians who are more familiar with the Old Testament than ordinary Jews suppose that Jews read the Old Testament the same way as born-again Christians. They should read *The Rationalist Bible* by Dennis Praeger, a Jew who has a favorable opinion of Christianity. Even prior to Rabbi Moses Maimonides (1138-1204), rabbinic Judaism had become rationalist in its understanding of the Old Testament. This is because unlike Jesus and the Apostles, the Rabbis did no miracles. Though believing in sorcery and Kabbalistic mysticism, rabbis do not read the Old Testament in the supernatural and apocalyptic way as did Jesus, the Apostles, the Old Testament Prophets, and Christians who believe the same. Rather, Rabbinic Judaism is a this-world religion, also being the root source of the secular interpretation of the Bible. Beginning with British Deists such as Anthony Collins (1676-1739), the higher criticism of the New Testament was taken over wholesale from Karite Jewish apologists.[41]

Notwithstanding that many today regard the Old Testament as a Jewish book, Jesus identified himself as David's Lord, the Heavenly King

---

[41] c.f. Richard H. Popkin, *Disputing Christianity: The 400-Year-Old Debate over Rabbi Isaac Ben Abraham of Troki's Classic Argument* (2007, Humanity Books, imprint of Prometheus Books)

of Old Testament Israel.[42] His disciples called him by the same title *Lord* as the Greek version of the Bible translated the Hebrew tetragrammaton translated as *Jehovah* or *Yahweh.* Recent scholars have pointed out the ancient Jews and even rabbinic Jews in the first centuries after Christ recognized two Heavenly powers which we would better understand as God the Father and his Christ, the Son through whom God created the earth.[43] John the Baptist prepared the way for Israel's Heavenly Lord, who he pointed to as being Jesus. It was centuries after Jesus that the rabbis developed their monotheistic polemic against the Trinity.

Medieval commentaries reveal that while Christians viewed Jews as sharing their roots, the Old Testament was also Christian history! Thus, the Old Testament served as a model for Christian kings and princes. However, it was this shift to viewing Jews as a race that has caused many Christians to suppose the Old Testament as a *Jewish* rather than Christian book. This also informed *the higher criticism of the Bible.* Aiding this was the fact that the German critics following Martin Luther misunderstood the Apostle Paul as confronting Jewish legalism. Paul however was addressing the end of the Old Covenant of Israel.

Looking forward to the millennium, the Reformers and especially the Puritans viewed the years of dominance of the church by the Papal antichrist as the Christian *Dark Ages* which would end in the promised *Enlightenment* when the Papal antichrist would be defeated. This was how premillennial English Protestants understood what is now called the *Enlightenment* before it was secularized as the leadership of the Enlightenment moved from Protestant England to Voltaire's Catholic France. Thus did Enlightenment historians such as Edward Gibbon (1737-1784) see Christianity as itself responsible for the decline of civilization as once existed in the classical civilizations of Greece and Rome. Following Gibbon, secularists began teaching the Christian centuries as being the Dark Ages between the classical pagans and the Renaissance.

THE REVIVAL OF RACISM BY THE THINKERS OF THE ENLIGHTENMENT

Thinkers of the Enlightenment such as Montesquieu (1689-1755) who admired the classical pagans also adopted the geographic-based racism of the pagan Greeks, applying it even to the lands of Europe.

---

[42] c.f. Matthew 22: 41-46

[43] c.f. Alan F. Segal, *Two Powers in Heaven: Early Rabbinic Reports about Christianity and Gnosticism* (2002, Brill; reprinted in 2011 by Baylor University Press) and popularized by the late Michael S. Heiser, academic editor for *Logos.*

According to Montesquieu, those living in southern Europe were too hot-blooded and emotional. Those living in northern Europe were too cold and reserved, while those living in France were temperamentally just right, *temperate!* These anthropological designations would only later be used to describe climate. However, a white/colored racism would soon replace the traditional Christian/Heathen division of mankind. This was a philosophical, scientific, and economic defense of slavery and colonialism at a time when the authority of *reason* was challenging the authority of the Bible. If Christian missionaries were holding back economic progress by defending the rights of slaves and natives, "reason" was on the side of good business and science.

As early as 1619, slaves (a word deriving from the blonde-headed Slavs) were no longer emancipated after converting to Christianity as formerly the case among British Christians, many of the English being themselves descendants of former slaves. This new shift however applied only to black slaves. Associating blackness with slavery dates from the teaching of the Curse of Ham taught in the Babylonian Talmud, being perhaps a Jewish apologetic against the scandalous Christian teaching of the equality in Christ between masters and slaves. Whatever the reason, this Curse of Ham teaching was used by Jewish and Islamic slave traders, then by Christian slave traders from the beginning of the African slave trade.

Having rejected biblical history, the major figures of the Enlightenment such as David Hume and Immanuel Kant, introduced the most vile and ugly racism ever conceived in the heart of man. Following the Scottish Primitivists, Voltaire taught that Africans descended from orangutans.[44] Kant, the darling of liberal Christians who may be unacquainted with his anthropology, gave instructions on how to beat black slaves. But Kant viewed native Americans as useless wild savages, the most degenerate of all the races. Racist teachings were introduced into America through Thomas Jefferson's *Notes on the State of Virginia* (1785). The men of the Enlightenment were divided as to whether the races were from different creations, a view known as polygenesis, or as having degenerated due to being scattering to inferior lands. Thus, Thomas Jefferson had to defend American wildlife as not being inferior to the Old World animals.

While those of the French Enlightenment such as Voltaire preferred polygenism, those in Germany such as Jean DeLuc's friend Johann Friedrich Blumenbach (1752-1840) who invented *Caucasian* as one of

---

[44] Race and the Enlightenment, A Reader (1997), Edited by Emmanuel Chukwudi Eze

his five basic races, explained the other races as due to the degeneration as the other races settled in their inferior lands. Blumenbach, who made fun of the account of Noah's Ark, hardly based his explanation on the Bible. Louis Agassiz, the father of the Ice Age, was not alone among nineteenth century scientists who believed that Christianity itself has spread from the white Caucasian race. While Charles Darwin famously saw the white races as the fittest for survival, at a time when slavery was being outlawed in Great Britain due to the work of the British Evangelicals, his mentor Charles Lyell was an apologist for Southern slavery!

THE BEGINNINGS OF ANTISEMITISM

Today, race-based *antisemitism* is falsely equated with anti-Judaism, the latter being a theological rather than racist judgment. While religious Jews of the New Testament were anti-Christian, the Apostles were anti-Jewish regarding the Jews who opposed Jesus. Neither were antisemitic or racist. It was only in the late eighteenth century that a supposed inferiority of Jews would be explained as being due to their "Semitic" homeland which they shared with the Arabs. This view developed due to J. D. Michaelis (1717-1791) teaching that the Hebrew language derived from Arabic, which he believed existed in purest form in Yemen.

Not wishing to see their nation as deriving from the obscure Gomer, son of Japheth, or any other nation as could be determined from the Bible, German nationalists who were inspired by the Neo-Pietists J.G. Hamann(1730-1788) and J.G. Herder (1744-1803) sought a homeland of the German race in some more German land, the most popular being in Persia were lived the ancient Aryans (from which the nation of Iran would be named).[45] Thus, Jews were seen as hindrance to German nationalism due to being "Semitic" and thus not German.

Shifting to the teachings of Darwin, German nationalists embraced the absurd view of supposing that Jews are a biological race. This meant that the many Jews who were atheists or had converted to Christianity remained Jews. It also meant that if one had Jewish ancestors, he or she still belonged to the Jewish "bloodline," a Darwinist and Nazi obsession deriving from animal breeding and the inheritance order of

---

[45] c.f. Tuska Benes, In Babel's Shadow: Language, Philology, and the Nation in Nineteenth Century Germany (2008, Wayne State University Press)

the Old World *ancien regime*. Sadly, even some Christians have come to view Jewishness as a matter of "bloodline."

THE APOGEE OF RACISM: THE BEGINNING OF THE TWENTIETH CENTURY

Combined with Social Darwinism, these teachings led to European nations not only looking down upon Jews but also on Slavs and the "inferior" European nations of southern Europe. This led to prejudices against American immigrants from southern Europe. Racism even divided the Germans from the English, leading to the great carnage of WWI.[46] This apogee of racism coincided with these nations' shift in intellectual authority from the Bible to Science.

THE MARXIST SOVIET EXPLOITATION OF WESTERN RACISM

Promoting the neo-pagan racism that derived from the teachings of German nationalists, Hitler and the Nazis viewed Marxists communists whose leadership included Jews, as their greatest enemies. They also viewed the Slavs of Poland and Russia as an inferior race who could be eliminated so as to provide more living space for Germans. This led to Nazis slaughtering Slavs even before they began the 'final solution' to the "Jewish problem." One can thus understand why to this day, Russians opposed the Nazis. In an era of excessive racism in the West, neither race nor ethnic identity was a barrier for leadership among the Marxist, atheistic as they may have been. During the Cold War, the Soviet Union exploited the existence of racism in America while courting the third world nations of Africa, Asia, and America, as they also exploited nationalistic movements in the same countries who were seeking independence from colonial powers.

This has caused some to wrongly suppose a concern for non-white peoples as being a Marxist agenda. But however much the Marxists exploited America's racism, the movement to eliminate racism in America and the West as also happened in the beginnings of Christianity, were Christians. The success of the Civil Rights movement was due to the Southern Christian Leadership Conference founded by Martin Luther King, Jr, and supported by evangelicals such as Billy Graham.

---

[46] c.f. Vernon Kellogg, Headquarters Nights: A Record of Conversations and Experiences at the Headquarters of the German Army in France and Belgium (1917, Leopold Classic Library)

The Marxist solution to racism is to destroy all history and culture and teach people to be content with whatever might be assigned to them by the state, a system commonly referred to as socialism. Under Marxist government, the rights of individuals are replaced by an elite party and usually a dictator who governs a totalitarian state on the principle of the ends justifying the means. Neither can Marxism provide a solution to the emptiness and despair as exists within the human soul or allow such liberties as exists in America due to the nation's Christian roots.

The answer to Marxism however is not the kind of market capitalism which results in the best government that money can buy. Rather, it was Karl Marx who popularized the word *capitalism,* a word which seems to have originally arose from shares of the heads (caps) of slaves in the charters of stock companies involved in the Mediterranean slave trade. Rather, it requires a voluntary Christianity, the eternal solution to racism, poverty, and all manner of injustice. The same Christian missionaries and revivalists who challenged slavery also challenged racism, eventually resulting in the emancipation of racial minorities. Ironically, the end of racism as a respectable view came from a reaction against the horrible Holocaust birthed due to viewing Jews as a race! As I will explain in the next chapter, the vile racist scientific prehistory of man continues to be taught to this day in academia. It is time to replace it by teaching how all mankind descends from one family who were saved by Noah's Ark.

36

UNDERSTANDING PREHISTORY THROUGH DIFFUSION AND MIGRATIONS

If the Industrial Revolution together with belief in progress inspired the theory of evolution, it was the well-known facts of European and ancient history that informed the beginnings of scientific archaeology as pioneered in the Scandinavian countries. The actual history of man knows nothing of unilinear evolution. Rather, social, political, economic, and technological changes are specific in their origins and spread to other lands through migrations, trade, missions, invasions, and neo-colonialism as was peaking in the latter years of the nineteenth century. The latter however can also bring civilizations to an end as resulted from the barbarian invasions of Rome and from European invasions of native civilizations in Mexico and Peru.

Material degeneration necessarily accompanies long distance migrations as migrating peoples get isolated from market economies. Thus, a shift to a less refined, more rustic style of living occurred with the first European settlers of America. Spiritual degeneration may also occur if those who embark on long and dangerous journeys are seeking to gain wealth by capturing slaves or acquiring land by eliminating native peoples. Those who venture to new land often include criminals seeking to escape punishment or else were sent to the new lands as punishment. But a few may be law-abiding citizens willing to endure great hardships to enjoy religious and other kinds of freedom while aiming to live peaceably with those who had migrated to the great undeveloped wildernesses long before them.

A similar material degradation happened immediately following the Flood. Due to the great loss of population and consequently specialization in the practical arts, those surviving the Flood faced a more rustic if also more peaceful style of living than existed in the materially refined Early Bronze Age. The art of the fast wheel, the making of painted and stone pottery, the construction of great stone buildings with amazing precision, and likely many other arts were temporarily or permanently lost because of the Flood though not important ones such as writing, spoked-wheel wagons, and building ships and boats.

The need to uphold unity and order which are essential for survival over long and difficult journeys explains the hierarchical societies of many native peoples, where obedience of subject peoples might also be enforced by torture. The widespread practice of cannibalism as found in the Pacific islands and the New World is likely the result of the long crossing of the Pacific Ocean. However pre-Christian Europe was not much different. Torture and slavery were common practices in the classic civilizations of Greece and Rome as in almost every pre-Christian nation, including Phoenicians and Celtics who also practiced human sacrifice.[47]

Various kinds of degeneration explain the spread of man as seen in the light of the Bible, but a view held only by a few anthropologists such as the Archbishop of Dublin Richard Whatley (1787-1863) and Lyell's Canadian friend John William Dawson (1820-1899). Whatley argued that there is no evidence that savages unaided had ever developed a less barbarous way of life. Thus, it followed that 'humanity must have originally existed in a state far superior to that of modern savages.' This is the very opposite of the view of unilinear evolution which has dominated post-colonial archaeology. But nineteenth and early twentieth archaeology explained changes in human societies as coming from migrations and from the diffusion of progress from advanced civilizations.

[47] c.f the atheist(!) Tom Holland, Dominion: How the Christian Revolution Remade the World (2019, Basic Books)

37

LIGHT SPREADING FROM THE ORIENT

Following the path of his Scandinavian predecessors, Oscar Montelius (1843-1921) developed the first relative chronology for European prehistory by seriating the artifacts he found in the museums of European nations. His seriation showed that the peoples of northern Europe had spread from the Orient, as the ancient lands of the Bible were then called. V Gordon Childe (1892-1957), who would further develop Montelius's prehistoric European chronology, would summarize his predecessor's reasoning in 5 propositions:

1. Civilization in the Orient is extremely ancient.
2. Civilization can be diffused.
3. Elements of civilization were in fact diffused from the Orient to Europe.
4. The diffusion of historically dated Oriental types provides a basis for bringing prehistoric Europe within the framework of historical chronology.
5. Prehistoric European cultures are poorer than contemporary European cultures means that civilization is later in Europe than in the East.

Montelius was the first to subdivide the Stone Age, Bronze Age, and Iron Age periods into epochs of Early, Middle, and Late using seriation of styles. He further subdivided these periods by a series of Roman numerals. William F. Albright would later extend these classifications into the Levant of the Ancient Near East.

Montelius's understanding of the light of civilization spreading from the Ancient Near East reflected the work of anthropologists of his day who saw civilization spreading by diffusion as might happen through contacts such as missionaries or trade. Following those who believed that civilization began in Egypt, Grafton Elliot Smith and W.J. Perry taught that the arts of embalming and pyramid building spread even to the Americas, though being increasingly of an inferior, degenerate style. The Americas however have far more similarities with the nations of Far East Asia.[48]

The idea of diffusion and migration didn't necessarily challenge unilinear evolution, but the ethnologist Friedrich Ratzel (1844-1904) identified the spreading of inventions and progress as being due to a specific and superior culture. Believing that certain inventions such as the bow and arrow are too complex to have been invented more than once Ratzel insisted that progressive development depended on the spread of inventions from superior nations like those of Germany. He aimed to find more living space for the German people, thereby influencing the rise of German imperialism. In doing so, he visited places where German immigrants had spread their superior German ways to the United States. His disciple Franz Boas (1858-1942) used Ratzel's concept of culture to develop the first prehistoric chronology for pre-Columbian America.

---

[48] Michael D. Coe, *The Maya* (1993, Thames and Hudson), Chapter 2

38

ARCHAEOLOGICAL *CULTURES*

Ratzel and Boas sought to understand archaeology by studying the migration of *cultures*. As early as 1851, Daniel Wilson, who was the first to use the word *prehistory,* had noted stylistic differences in artifacts belonging to Iron Age sites in Scotland and those in Scandinavia. Different artifact styles may reflect different peoples rather than stages of development as presumed by the unilinear evolution of culture. The term that came to be used for a distinct assemblage of material remains believed associated with distinct people groups is that of an archaeological *culture,* as the case of the Khirbet Kerak culture that I mentioned above which spreads from Mount Ararat into Israel.

An archaeological culture is the extension to the material remains of ancient people of a term formerly applied to the shared language, values, manners, religion, government, and laws of a particular class or nation. J. G. Herder (1754-1803), who along with Immanuel Kant and J.G. Blumenbach (1752-1840) founded competitive schools of scientific anthropology in Germany, coined the word *culture*, a term formerly used in agriculture, applying it to the distinctive literature, religion, and traditions of those sharing a common language. While the racism of Kant and Blumenbach was based on geography, racism could also be based on language and culture. Herder's theory gave rise to the nationalism of the Romantic era that challenged the internationalism of the Enlightenment such as Frederick the Great had promoted in Prussia.

Herder had particularly in mind his German language then spoken in numerous independent principalities across Central Europe. Following Napoleon's invasion of Prussia and a famous 1808 speech by J. G. Fichte, German nationalists sought to establish a more prosperous and powerful German nation by uniting the German speaking states. Germany would thus become the first geopolitical nation created on the basis of a common language since ancient Babel, though it would inspire many other groups to do the same either through peace treaties, post-colonial independence movements, terrorism, or war.

Inspiring the creation of a nation based on language requires those speaking such the language to have shared a glorious history. German nationalists faced a problem due to the fact that the written German language was the creation of Luther's sixteenth century translation of the Bible. Hence, the German speaking peoples had no written annals by which to teach the members of the new nation a shared and glorious history. German nationalists however hoped to find the origin and ancient history of those speaking the German language through comparative philology and the study of German culture, including its earlier pagan religion and orally transmitted folk tales. Thus the linguists known as the Brothers Grimm collected and publish German folk tales.

The idea that history can be reliably preserved through oral means went back to Isaac Newton's teaching about Egypt and Israel's preservation of history prior to the time of David and Solomon when he believed that writing was invented. Newton was perhaps influenced by the rabbis who claimed that the teachings of the Talmud had been passed down orally from Moses. These German nationalists claimed that the same was true of the Zend Avesta, the religious writings of the ancient Iranians, as well as the Vedas of India. There is however no evidence that any of these Eastern religious texts were written down if they even existed earlier than the Christian centuries. Not surprisingly, the German higher critics supposed the same processes in the composition of the New Testament, though in truth composed by or from eyewitness evidence.

Poems and songs are useful for teaching and preserving sagas and traditions in pre-literate societies. But maintaining schools devoted to teaching poems and songs by oral transmission requires the stability of a priestly class which, owing to wars, changes in governments, social and economic positions, can hardly be maintained over such long periods. The reliability of oral transmission can be judged by the wide variations in the accounts of the Flood preserved by the various nations. The oral stories behind the Vedas probably date from Cyrus's conquest of India, while the caste system of the Hindu religion reflects the class society of the ancient Greeks. These Eastern religions also appropriated reincarnation from the teachings of Plato. This suggests that Hinduism dates only from the conquest of India by Alexander the Great. The tendency of this religion to seek and claim ancient roots also reflects the same Greek culture. My scholarly Indian friends tell me that the oldest temples in India are not those of the Hindus, but those of the Buddhists!

Searching for the origins of nations through language requires the disciplines of philology and comparative linguistics. The British philologist William Jones (1746-1794) helped launch such studies by noting the Indian Sanskrit language's close kinship with European languages including Greek, Latin, German, and Celtic. Based on the table of nations in Genesis 10, Jones referred to these as *Japhetic* languages. But not wishing to see the German nation as coming from the obscure Gomer, son of Japheth, son of Noah, and from a religion and literature of Semitic peoples, German nationalists preferred to find the original homeland of those who they called Indo-Germans either in Iran, India, in the classical civilization of Greece, or even in western Central Europe north of the Alps.[49]

One ancient culture speaking an Indo-German language that German nationalists believed reflected the martial culture of which Prussian Germans were very proud was recognized from words pertaining to horsemanship used by the ancient Aryans [Iranians] of Persia who they believed *to* be the same as those who invaded the dark-skinned peoples of India as mentioned in the Vedas. The Sanskrit language in which these Vedas were written is essentially Old Persian which was carried into India by the invasion of King Cyrus who is mentioned in the Bible. But German nationalists aimed to avoid the "Semitic" Bible and discover this history entirely from pagan sources, real or imagined. Thie is why the term *Aryan* replaced *Indo-German*. German nationalists even hoped to find the roots of German Christianity in the Caucasus, Assyria, or some other "non-Semitic" land!

Cultures such as those speaking the German language could perhaps be associated with distinctive archaeological spreads. Identifying the spread of archaeological cultures with those speaking the German language was an idea based on the German tribes who invaded Western Europe following the fall of the Roman Empire. Using the swastika found on ceramics in the spread of the Aryan race, Gustaf Kossinna (1858-1931) sought the homeland of the Aryan races north of the Alps, all of which Hitler aimed to recover for German settlers. Kossinna attacked colleagues who sought the roots of German civilization in the Orient or on one of classical nations of southern Europe. V Gordon Childe however was sympathetic to the Marxists of

---

[49] Tuska Benes, In Babel's Shadow: Language, Philology, and the Nation in Nineteenth-Century Germany (2008, Wayne State University Press)

Russia who preferred to see civilization radiating from the Slavic lands such as those nearer the Caucasus.

The Caucasian origins of white people as promoted in Europe and America by Ice Age pioneer Louis Agassiz was popularized in the nineteenth century by the scientifically acclaimed *American school of ethnology.* The American school was based on a wide collection of skulls by the physician Samuel George Morton (1799-1851) but understood according to the teachings of his followers Josiah C Nott and George Gliddon. The American school claimed there existed measurable differences in the skulls of races which indicated differences in intelligence. Arising prior to the teaching of evolution, the racism of the American school was based on the teachings of Isaac LaPeyerre as mentioned in Chapter 35. These ethnologists lamented how a literal reading of the Bible, where all races are seen as having come from Noah's family, was holding back the progress of science. LaPeyerre was compared to Galileo as a persecuted forerunner of science. Though the teachings of LaPeyerre were soon to be replaced by the racist teachings of Darwin, this school's view of the importance of the size and shape of skulls continues to inform modern anthropology's view of the antiquity and evolution of man. Even more did it suit the new Aryan view of the origins of white people.

Leaders of the American school of ethnology translated the writings of the French aristocrat Arthur de Gobineau (1816-1882) whose racist teachings were aimed at supposed differences in Europeans as could be seen in the class system of Europe's *ancien regime*. The German born philologist Max Mueller (1823-1900) taught the Aryan theory in England even if he regretted some of the racist implications. These teachings were much responsible for the invasions, mistreatment, and mass executions of civilians during both World Wars. Both diffusion and racist inspired migrations dominated late nineteenth and early twentieth century anthropology and prehistoric archaeology. But whether racist, nationalistic, or evolutionist, archaeological prehistory was written to accord with the values and views prevalent in the Neo-Colonial powers of the late nineteenth and early twentieth centuries.

After the fall of Hitler, the Aryan languages were renamed Indo-European, as originally suggested by Thomas Young. But philologists and archaeologists continue to search for the original homeland of those speaking the Proto-Indo-European language (PIE), most often finding it either in ancient Anatolia or the Caucasus. These aren't so distant from Mount Ararat where exists the ancient homeland of those speaking all languages.

39

THE POST-COLONIAL EMBRACE OF NEOEVOLUTION

The archaeological migrations by which I traced the worldwide migrations from Mount Ararat were based on the work of archaeologists such as Montelius and V Gordon Childe in Europe and Franz Boas in the Americas who worked from the late nineteenth to the middle of the twentieth century. So are the prehistoric chronologies still being used across the world. Unfortunately, by the nineteen sixties studies of worldwide migrations together with an integrated historical understandings of mankind's prehistory were no longer being studied. Rather, studies of migrations were replaced by the processual archaeologies which presume that civilizations throughout the world developed independently of the Ancient Near East where the Bible reveals as being the birthplace of all mankind. There has since been a reaction against processional archaeology. However, due to the influence of Darwin, prominence is now given to out-of-Africa theories. It seems that ancient peoples forgot how to migrate once they left Africa.

There were however reasons for this shift. One was the cultural imperialism and racism as I have just discussed that so informed the archaeology of this period, the most dramatic being the archaeology of Aryan invasions based on German nationalism which prepared the way for Hitler. There were other abuses including racist inspired archaeology in such places as America, Africa, and Australia. Montelius had avoided the tendency to look at migrations as invasions while seeing the diffusion of culture into Europe radiating from the Orient. Though combining Kossina's concept of culture with Montelius's relative chronology to further develop European prehistory, due to being a Marxist, V Gordon Childe did not champion one race above another. Nonetheless, he had begun his work in archaeology by trying to understand the origins of the Aryan peoples. Archaeologists who did not wish to see the civilization of Europe as having come from the lands of the Bible used Childe as a reason to attack the study of archaeology in the light of diffusion or migration.

The reason for such confusion in archaeological prehistory was due to not recognizing that most prehistoric peoples lived before the Flood. Archaeologists must however work with whatever material remains survived from these ancient civilizations. Thus, there was a turn to neo-evolution which differed from the older evolution where civilization might have been advanced by special gifted individuals, nations, or spiritual sources. According to neo-evolution, mankind's future is fully determined by his physical environment. This meant rejecting historical sources for the rise of civilizations as the case of peoples or influences spreading from the lands and nations of the Bible.

The strongest case for agriculture and civilization having developed independently of the Ancient Near East was in America where domesticated plants were almost entirely different from those in the Old World. Civilization in America was thus dated from the Paleolithic cultures believed to have crossed from the Old World by way of a temporary land bridge arising from falling oceans during "the last glacial maximum." The idea that native Americans had crossed from Asia via a land bridge is however far older than the Ice Age theory. It dates to the speculations of the earliest European explorers.

Europeans did not understand that Far East Asians had long been superior mariners. Wooden boats do not long survive, but archaeologists should have learned from anthropologists that the most "primitive" peoples knew how to fish and build boats from the time that Europeans discovered them. Should not the children of Noah have retained maritime skills? The earliest settlements being along the Pacific coast makes it clear that that America was first settled by fishermen.

Before and after the Flood, the first American settlements were in Ecuador.[50] The name *Ecuador* derives from the equator, which runs through this country. Experienced sailors understand the danger of ships that have lost their direction of sailing in circles. Thus, the safest way for lost mariners to avoid sailing in circles is to follow the sun. No matter where in the Pacific one begins his voyage, by following the sun,

---

[50] Betty J. Meggers, Clifford Evans, and Emilio Estrado, *Early Formative Period in America: the Valdiva and Machalio Phases* (1965)

a ship will eventually arrive in Ecuador. The Formative Culture for native American civilization thus begins in Ecuador.[51]

This mistaken belief of native American civilizations' supposed independence of Old World Archaeology was not the only reason that Americans took the lead in what by the sixties was being called the *New Archaeology.* This trend was led by archaeologists inspired by the invention of radiocarbon dating, by statistical analysis of artifacts, by the scientific work of paleobotanists who studied ancient diets recovered from archaeological sites, and using computers in the service of archaeology! For some, archaeology seemed on the verge of becoming a "science" instead of just belonging to the humanities. Indeed, scientific methods are useful for archaeology. No archaeological project needs new technologies and methodologies more than the scientists who study the region of Mount Ararat.

Advocates of what was called the *New Archaeology* supposed that archaeological sites and rise of civilizations might be explained according to General System Theory. This was in line with the scientism that the American social sciences had been promoting from the end of the nineteenth century which aimed to explain America's exceptionalism as due to the nation's leadership in science.[52] As a new nation, America lacked the long history and the numerous historical places and markers that informed Old World archaeology.

As a systems engineer and pioneer of the world's first distributed digital and fiber optic networks, General Systems Theory was a useful tool for my profession. I used a systems understanding to automate the world's most complex infrastructure networks. Subsequently, I employed a systems methodology to search for archaeological evidence of Noah's Flood. However, I knew that the key to developing innovative solutions was understanding the history of telecommunications technology and infrastructure *history.* I likewise learned archaeology and geology from the history of these fields, the same as I am sharing here with you. Far from developing technology by focusing on technology, I head to learn the specific needs of the various kinds of rural and urban areas for which I was designing these systems. I had also to become familiar with the operations of various kinds of companies and government agencies who operate telephone networks. Thus, to excel in engineering, one must be interested in human beings

---

[51] James A. Ford, A Comparison of the Formative Cultures of America (1969)
[52] c.f. Dorothy Ross, *The Origins of American Social Science* (1991, Cambridge University Press)

and history, the opposite of the stereotype of engineers as technical nerds! How astounded I was to learn that archaeologists supposed they could understand civilizations apart from the light of their specific histories.

Such pretentious systems thinking that has been employed to explain the rise and collapse of the great pre-Flood civilizations. Explanations using abstractions such as *complexity* and *collapse* merely restate the problem. Thinking in terms of abstractions masks useless gobbledygook. As I shared with the President of the *American Schools for Overseas Research,* this turns archaeologists into glorified technicians.

Archaeology however was not the only academic field traditionally belonging to the humanities that scientism was destroying. *Scientification* based on subjective abstractions so as to follow fashionable trends was undermining higher education in America. Such teachings are often accompanied by the idea of a progressive scientific elite who should run society. Sad to admit, as an honor-class engineering student in in the late sixties, the idea of a scientific elite once appealed to me.

Ironically, that means that the earlier racist inspired migrations upon which prehistoric chronologies are based continue to interpret archaeological sites in the processual understanding that replaced archaeological prehistory. Migrations however are a fact of history and need not be interpreted in racist ways. Recognizing that the people of all nations belonged to the single family who less than five thousand years ago began their migration to all parts of the earth. Considering that Noah had three daughter-in-laws who descended neither from Noah or his wife, we should not be surprised that the peoples of the world reflect those three genetic divisions, while those living closest to Mount Ararat reflect them all.

40

Establishing Prehistoric Chronologies for the Ancient Near East

As Montelius and Childe sought to establish a relative chronology for prehistoric Europe, scholars and archaeologists working in the lands of the Bible had been seeking historical and prehistorical chronologies for what Childe called the *Most Ancient East*, now referred to as the *Ancient Near East* (ANE). The Ancient Near East is essentially that of the lands mentioned in the Bible. *Middle East* is the location of a modern military command. Unlike barbarian Europe as well as the classic civilizations of Greece and Rome, literacy had long been present in the lands of the Bible. As Western nations obtained access to lands of the Ancient Near East beginning with Napoleon's invasion of Egypt, archaeologists found monuments, inscriptions, and clay texts with the names of the same Assyrian and Babylonian kings known from the Bible, some even referencing kings of Israel. Explorers found trilingual inscriptions as those on Egypt's Rosetta Stone they learned to read Coptic and hieratic Egyptian. Within sight of Mount Ararat in Kurdish Iran, they discovered the Behistun inscription by which they learned to read Persian, Elamite, and Akkadian. From these, they learned to read the Sumerian and Hurrian languages.

Using names and king lists written on monuments and ancient papyrus from Egypt and on clay tablets in Mesopotamia, scholars were able to construct more extensive king lists by which to establish historical chronologies from even more ancient times. Some of these texts mentioned a Flood like the one in the Bible, including kings who lived beyond ordinary lifespans, the earliest of whom lived before the Flood. Working from these, scholars established a historical chronology for the earliest dynasties back to about 3100 BC in Egypt and 2800BC in

Mesopotamia. However due to gaps in these king lists, the earliest reliable historical chronology dates only from the beginnings of the Neo-Assyrian empires in the late second millennium BC. The much-discussed chronology of Egypt, including the proposed dates for the Exodus, is no exception.

Flinders Petrie (1843-1942), who saw civilization in Egypt as coming from an invasion of Africa by a superior race from the North, was chiefly responsible both for developing an Ancient Near Eastern historical chronology and for connecting Montelius's prehistoric European chronology to the Ancient Near East. Working with Arthur Evans who I mentioned in Chapter 31, he connected his historical chronology of Egypt to those of the Aegean, while Mortillet connected the Iron Age Hallstatt and La Tene cultures to the historical Celts who would eventually inhabit Britain. For the prehistoric eras of the Ancient Near East however the archaeologists could seek only relative chronologies. Following Montellius, Petrie sought to accomplish this through a seriation of pottery based on his proposal for a supposed development of pottery styles. Petrie's supposed periods of earliest Egyptian history should not be taken seriously due to the lack of stratified sites. Thus Petrie began working in Palestine, where he could link prehistoric levels to those in Egypt.

41

How the Flood Confused Ancient Near Eastern Chronology

Digging through the ruins of ancient Nineveh and in sites extending into lower Mesopotamia, archaeologists found a series of pottery which they imagined as having been produced in a succession of prehistoric eras. They named these for various sites where the pottery had first been identified. These ranged from Halaf, Hassuna, and Samarra nearer the top of the Fertile Crescent to distinctive kinds of pottery found at the sites of Ubaid, Uruk, and Jemdet Nasser in the south. The latter are followed by Early Dynasty remains, the first historical period of the Sumerians which according to the current historical chronology began around 2800BC.

As I mentioned above, aside from sites in the highlands, Early Bronze Age ruins found under tels in the great river valleys were too deeply buried by the Flood for easy excavation. The same is necessarily true of everything supposed earlier. In comparison to the upper layers where were found the historically dated Sumerian and Akkadian remains, these deeper remains are massive. It is because these are the remains of what appears a densely populated, highly developed, and materially refined civilization that existed in Mesopotamia prior to the Flood. Many will be surprised to learn that the remains from the great Empire of Sargon the Great, reckoned as being the high point of Ancient Near Eastern civilization, are comparatively meager but near the surface and thus easy for archaeologists to recover.[53] The population of the Ancient Near East had profoundly dwindled due to what was then a recent Flood.

Some have wondered why Noah did not re-use the wood from the Ark. Dismantling the Ark and carrying these heavy timbers down this

---

[53] Harriet Crawford, *Sumer and the Sumerians* (1991, Cambridge University Press), p. 16

high mountain would have been an unnecessary burden because there existed plenty of worked wood from the dwellings of those who perished in the Flood. This wood however soon rotted and burned, causing archaeologists to suppose those buried in these sites were cremated, though most of such sites are only partially "cremated." Pottery, on the other hand, is more durable. It seems that the series of beautiful pottery including that found at Halaf, Hassunna, and Samarra were produced during the refined preFlood civilization but mined by the survivors of the Flood as they traveled from the land of Arrata along the Tigris and Euphrates Rivers. This pottery spreads in the same general area as that of Late Uruk which, as I mentioned above, is likely the site of the Tower of Babel. Contrary to the present chronology, Late Uruk which follows the Flood is later than Jemdet Nasser, which is one of those cities destroyed by the Flood.

Post-Flood remains seem to begin in what is now regarded as Early Dynasty IIIB remains. Ruins beneath the Early Dynasty IIIB remains are for the most part contemporaneous and dated to the antediluvian era. It now seems possible that none of these ruins including those buried by the Flood date much earlier than about 4000BC, suggesting than mankind may be no holder than about 6,000 years.

The same is true in Egypt. The Pyramids built during the earlier refined but violent preFlood civilization were likely for protecting treasures. This explains why the Egyptians had no memory of building the Pyramids which these historical Egyptians remodeled as tombs for Egyptian kings. Due to the Flood, the Archaic Egyptian language for which the hieroglyphic script was invented has never been deciphered, a matter that popular histories of ancient Egypt seem blissfully unaware. Neither are they aware that Egypt's Old Kingdom, Middle Kingdom, and New Kingdom eras are modern inventions imposed on the history of ancient Egypt. All these Egyptian dynasties however date from after the Flood. Evidence of the First Dynasty of Egypt was found at the site of Khirbet Kerak, as we might expect due to the historic Egyptians having migrated from the ancient wooden ship that I visited on Mount Ararat.

42

NEOLITHIC REVOLUTIONS OCCURRING AT THE END OF THE ICE AGE

We have next to consider the prehistoric chronology for the beginning of the Neolithic in the Ancient Near East which, as in every place else in the world, remains anchored to Lyell's magical 12,000 bp 'end of the Ice Age.' In 1914, the Egyptologist James Breasted (1865-1935), who founded the Oriental Institute of Chicago, coined the term *Fertile Crescent* to describe the crescent shaped region running from Egypt in the south to Mesopotamia in the East where he judged to be the place where agriculture and civilization began. In 1935, V Gordon Childe put forth the notion that successive agriculture and urban revolutions had occurred in the Fertile Crescent immediately following the end of the Ice Age. As a Marxist, Childe thought in terms of *revolutions*. According to Childe, the first revolution was when mankind suddenly changed from EpiPaleolithic hunter gatherers to Neolithic agriculturalists. His second revolution occurred after another few thousand years when Neolithic agriculturalist villagers learned to make pottery, forge metal, and build the world's first urban civilization where they would invent stamped seals and writing, these being chiefly for the purposes of accounting for their stores and commercial transactions.

Why and how did Childe suppose that after hunting and scavenging for hundreds of thousands of years, mankind should suddenly recognize the advantages of cultivating the grasses he was reaping and herding the flocks of wild sheep and goats that he followed? Following an idea put forth by Raphael Pumpelly in 1908, Childe suggested that it was because the Ancient Near East was becoming more arid during the colder Younger Dryas era supposed to have followed the Ice Age. This caused hunter-gatherers to spend more time at oases near the springs of rivers where plants and animals would be concentrated. The oases turned into villages where agriculture would begin. I suppose these oases could be seen as little gardens of Eden. Unlike the sons of Adam, it appears to have taken evolutionary man another two thousand years to recognize that the wild grains they were gathering were becoming more productive and the advantages that could be gained by sowing some of the wild grains.

Dorothy Garrod (1892-1968) managed to link the Paleolithic sites in Europe with the Neolithic sites in the Ancient Near East by exploring Neanderthal burials found in Gibraltar on the coast of Spain, in cave shelters in the foothills of the Zagros Mountains, and on Mount Carmel near the Mediterranean coast of Palestine. As was the case in Europe, she found the Near Eastern Paleolithic sites in rock shelters and caves on the sides of mountains. These are just the places where hunters might want to stay when away from their homes and villages. She connected these Epipaleolithic burials in the Schuqba Cave on Mount Carmel to nearby Wadi Natuf, where she discovered stone sickles and what has become accepted as the beginning of world agriculture.

The excessive antiquity of prehistoric man in the Ancient Near East due to the magical 12,000 bp 'end of the Ice Age' vastly stretched the antiquity of the ANE cities. Together with the fact that Charles Lyell taught archaeologists to equate depth-of-deposit with time, this greatly impacted the chronology of individual sites. Consider what this means if a 200-year-old city became entirely covered by the deposits of the Flood of a single year, thus creating an Ancient Near Eastern tel.

To make sense of these artificially created long-period sites, archaeologists excessively sub-divide the prehistoric chronological levels of sites that have neither been destroyed nor substantially rebuilt. Such attention to artificial periods averts attention from what was likely to have been a brief but rapid development of these sites and cities prior to the Flood. This is true of prehistoric archaeological sites all over the world but especially in the Ancient Near East as was the case of those searching for the world's oldest city.

43

DISCOVERING THE 'WORLD'S OLDEST CITY'

Robert Braidwood (1907-2003) began working with Breasted in the foothills of the Zagros Mountains where he believed that he would discover the world's oldest city. Dame Kathleen Kenyon (1906-1978) then working in Palestine at the site of ancient Jericho had the same aim. Being near the center of the Fertile Crescent and the world's oldest civilization in Mesopotamia, it was widely supposed that Braidwood's site of Jarmo was the most likely candidate for the world's first city.

Following John Garstang who began the excavation of Jericho, Kenyon reasonably determined that the spring coming from the bedrock that created the Jericho oasis was why Jericho became an early settlement. She found at the site evidence of the same kind of agriculture culture as Dorothy Garrod had found at Wadi-Natuf. This put the beginnings of the first temporary village at Jericho to perhaps 10,000 BC. Beginning around 8500BC according to the current chronology which Kenyon called the Pre-Pottery Neolithic A period (PPNA) were the first houses and a massive stone wall. She reckoned that the great defensive tower built around the spring and the finely plastered houses must have dated from the same time. In the following PPNB period, the occupants built rectilinear houses and learned the art of plastered skulls.

According to Kenyon, everything wonderfully advanced about the city including plastered walls as fine as that in some modern hospitals date to Jericho's initial phase and long before(!) Jericho's pottery. However, according to archaeological convention, pottery was the one thing that could establish the date of the city relative to other Ancient Near Eastern sites and would have therefore greatly lowered the date of the early stages of Jericho. In fact, there were plentiful amounts of pottery on the floors of these finely built houses, but Kenyon attributed every one of these pots to intrusions from above dug by later occupants! According to Kenyon, the first to use pottery in Jericho must have lived in tents because Kenyon could not find their houses. They left no other trace of their occupancy for thousands of years until the end of the Early Bronze Age when the city was suddenly abandoned.

Interestingly, Kenyon's famous stratigraphic drawing (Fig 7, Chapter 6) of the long trench which she dug into the Tel es-Sultan mound of ancient Jericho shows 'water-laid silt' lying just above the Early Bronze Age city.[54] The very next occupants of the abandoned Early Bronze Age city brought with them the Khirbet Kerak pottery that streams into southern Palestine from the mountains of Ararat!

William F. Albright, who created the archaeological chronology of the ancient lands of the Bible, and also his assistant George Wright noted Kenyon's tendency to separate pottery from strata at her other sites such as Tel Megiddo. Kenyon's mentor Sir Mortimer Wheeler knew no place on earth where there existed more sins in the name of archaeology than in Palestine. Wheeler may have had his student in mind. Archaeologists lamented the fact that Kenyon never published her archaeological excavation of Jericho. But something happened that caused criticism of Kenyon to cease and for archaeologists working in the Levant to use Kenyon's Jericho prehistoric chronology to date their own sites. This was the first archaeological application of the newly invented radiocarbon dating.

I explained in Chapter 4 why artifacts are rarely used for dating prehistoric sites. Organic artifacts such as worked wood, carved antlers, or leather rarely last for thousands of years. Thus, amorphous remains such as charcoal are more often used for radiocarbon dating. The interesting thing about Jericho is that it was also the center of the bitumen traded associated with the tar pits of the Dead Sea. Bitumen of course is a source of old radiocarbon. It also serves as fuel. Whether from natural or man-made fires, the wood of ancient Jericho was likely acquainted with smoke from bitumen fires. It is thus, not surprising that some of the charcoal from Jericho dates very early. But such is our world's confidence in scientific dating that to the dismay of Robert Braidwood, Kenyon's Jericho suddenly became the world's oldest city.

---

[54] Kathleen Kenyon, *Digging Up Jericho (1954),* Fig 4

44

THE WORLD'S "OLDEST" TEMPLE

Kenyon's work at Jericho was also used to date what is currently recognized as the "world's oldest temple." This is Gobekli Tepe in southeastern Turkey, a site excavated by the late Klaus Schmidt. Schmidt's method of dating was much like Kenyon's, that is, based on what he did *not* find at Gobekli Tepe, which is pottery. If someone does find pottery at this site, might it not be judged as being intrusive from a later era as was the case of Kenyon's "pre-pottery" Jericho? Thus, Schmidt dated this site to Kenyon's Pre-Pottery Neolithic Age (PPNA), which is now supposed as being between 9600-8,000 BC.

Klaus Schmidt has confirmed the PPNA dating of Gobekli Tepe with several radiocarbon dates, some from organic material clinging to these pillars, all which date from 9676-6749BC. I explained such radiocarbon dating from amorphous material in Chapter 4. But, as in the case of Kenyon's Jericho, nothing brings more attention, fame, and funding to a site more than finding the "oldest" site of something connected with humans, whether it be supposed ancestors, dwellings, cities, or temples. This has definitely been the case for the widely acclaimed Gobekli Tepe.

This site is easy to understand. This ancient quarry and workshop were turned into a burial mound and "temple" when the swift waters of Noah's Flood swept topsoil from the highlands of Eastern Turkey38 towards the Euphrates River, turning this antediluvian workshop into a large mound. Due to having been entirely covered, the American archaeologist who first surveyed this area supposed there existed here a Paleolithic camp. This was due to having found "Paleolithic" artifacts which very likely date from the same era as this "Neolithic" site. The archaeologists and those who study human prehistory can only see by the light of the current scientific narrative of the human past, but I am using a light which explains this puzzling site in a far simpler way.

What seems to have made Gobekli Tepe a candidate for the world's oldest "temple" is are the lack of evidence (i.e. pottery) for dwelling activities such as cooking and eating, combined with the presence of puzzling stone pillars and animal statuettes, some of which were in the process of being carved when this factory was ended by the Flood. Similar animal carvings are found across far Eastern Turkey including some that I have visited near Mount Ararat which would have better survived burial by topsoil washed down by the Flood. We may now know from whence these come; from sites like this if not from Gobekli Tepe itself.

The remote location of Gobekli Tepe away from sources of water which would be needed for permanent residence, combined with the fine limestone on the hill in where this site was founds points to Gobekli Tepe as being the quarry for these statuettes and limestone pillars. Some are found in an incomplete state, as the case of numerous other factories and stores at sites ended by the same Flood that buried this remarkable site.

The only thing of which we can be certain that religious interpretations of archaeological sites tell us concerns the minds and thinking of nineteenth century anthropologists and how they influenced archaeologists. The sociologist Auguste Comte (1798-1857) and especially the anthropologist Edward Tylor (1823 -1917) imagined an excessively religious prehistoric world. This served to blind archaeologists to the manifest evidence of the Flood as I explained in Chapter 8 above. Emile Durkheim (1858-1917) taught them that religion is something that can be understood scientifically!

I cannot decide which did more to blind our modern world to the truth concerning origins. Was it their aim to explain everything according to naturalistic science? Or was it due to the religious imaginations of scientists? Absent relevant texts, religious interpretations should be banned from the science of archaeology! Whatever the case, archaeologists will no longer be able to look at prehistoric burials without thinking about the Flood.

PART V

WHY THE RECENT DISCOVERY ON MOUNT ARARAT ARE LIKELY THE REMAINS OF NOAH'S ARK

45

The Recent Discovery

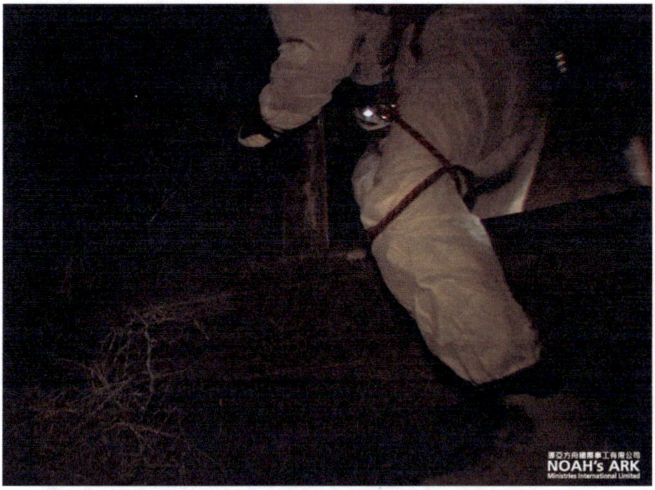

Fig 27 - Parasut climbs through an internal passage of the buried Ark

In May 2008, a small team of licensed professional mountain climbers led by my friend Ahmet Ertugrul who the world knows as *Parasut* uncovered an ancient wooden structure buried under volcanic rock and ice near the top of Mount Ararat. Parasut's search was sponsored by the Hong Kong-based media ministry *Noah's Ark Ministries International* (NAMI). NAMI immediately sent veteran British army trainer Panda Lee as an independent witness (Fig 1, Chapter 1). In 2009, a team led by Man-fai Yuen, the founder of NAMI, visited the discovery. All this was coordinated by Hong Kong public television reporter and documentary producer Clara Wei who immediately sought archaeological and scientific involvement. In April 2010, Clara's media company and NAMI, assisted by Netherlands-based ArkInsight Foundation, arranged for the discovery to be announced to the world.

According to the natives of the mountain who have held this secret from times immemorial, these remains are part of an ancient ship that once rested on the saddle of greater Mount Ararat which their ancestors have always believed to be Noah's Ark. According to the tradition of the family who has ruled this mountain from ancient times, they had been commissioned by God to protect the remains until such time as God would determine to reveal their secret to the world.

How do I know of this? In December 2015, accompanied by a team of scientists and videographers led by Clara Wei, I was honored to be the first outsider with whom these natives of the mountain ever agreed to meet.

Fig 28 - The author with the oldest living native of Mount

Mount Ararat is the highest mountain in the lands of the Bible. In the years following the Flood, snow and ice began to build on the top of this mountain which is almost 17,000 feet (5100 meters) above sea level. At some point, the ship became completely buried under the glacier that still covers the top of this mountain. Though there are reports of visits to the Ark from before the time of Jesus, as the glacier increased in size and depth, visits became more difficult. Through most of the Christian centuries, the site was protected by the Saint James monastery located on the north side of the mountain.

In the same 1840 earthquake that destroyed the Saint James monastery, the glacier enclosing this ancient ship slid down the south side of the mountain carrying with it the frozen ship. The long and narrow Ark rested on a ledge further down the mountain from which broke two major sections that slid further down the mountain. All three sections were covered by volcanic rock and glacier ice loosened by the earthquake. These sections now rest in three locations which the natives have designated as Sites A, B, and C. Rocks which continue to fall from the mountain add additional covering causing the Ark to become ever more deeply buried.

As one might expect, wooden boards from the wreckage were scattered on the highest parts of the mountain which are far above the tree line and well above where any structures exist whether for humans or animals. Shepherds do not graze their flocks so high on this dangerous mountain where exists only snow, ice, and rocks. Over the last few centuries, pieces of scattered wood have been seen on the mountain. These boards have not gotten much attention because Ark Searchers are not looking for scattered wood. They mostly search for the Ark on the northern and western sides of the mountain where the natives have directed them because it is far from the wreckage of the Ark. However, the French explorer Fernand Navarra (1915-1992) may have found some of this wood. As I can determine from the narrative of his journey, the site viewed at a distance in 1943 by the American serviceman Ed Davis appears the same as I visited in 2014.

More interesting are the artifacts spilled on the mountain at the time of the earthquake which the natives have collected and preserved in private collections in the nations surrounding Mount Ararat. The boundaries of these nations have changed since 1840. While I viewed some of these collections in public and private museums, this homeland of all the earth's people also preserves the world's oldest antiques, including spindles and looms for making textiles that may not be so different from those preserved on the Ark.

Fortunately, there remain well-crafted leather, wood, and bone artifacts still resting inside the Ark on the mountain. These will help archaeologists provenance the scattered artifacts and determine whether some of them may have been among those spilled from the wreckage on the mountain. The archaeology and history of the homeland of all the earth's people is at last beginning to obtain the attention it deserves.

46

THE DISCOVERY'S SCIENTIFIC STATUS

As I mentioned in Chapter 1, as soon as the explorers made their discovery, they sought an archaeological investigation. This got the attention of archaeologists because would be “Indiana Jones” adventurers seeking glory and money avoid archaeologists as much as they avoid questions in public forums. The archaeologists were also impressed with the photos and videos which the explorers had published. However, the archaeological community pulled back following the charge that the discovery might be a hoax. Archaeologists are busy with their current projects and do not wish to get involve with controversies. Becoming associated with a hoax will ruin the career of an archaeologist. Until this discovery, just investigating "Noah's Ark" would have been sufficient to destroy an archaeologist's reputation.

One might compare this world-changing discovery to the Dead Sea Scrolls, which likewise were not discovered by professional archaeologists. As the case of this discovery, there had long been rumors of the existence of these scrolls. For ten years after their discovery, the Dead Sea Scrolls were regarded as forgeries. What this means is that those who discover some of the world’s most remarkable archaeological treasures may not understand applicable archaeological procedures and rules pertaining to antiquities. Regardless NAMI’s intention to follow every law and regulation, it was no different with this discovery. Fortunately, with the help of the authorities, the discoverers have demonstrated a rapid learning curve. To protect the site from unauthorized visitors and preserve it for future archaeological study, they have avoided further visits and announcements concerning these exciting sites.

Archaeological interest revived after ASOR Trustee Sheila Bishop and I visited the ancient wooden structure on November 9, 2014. Today, it is the custom of nations that archaeological licenses be granted to universities and institutions in the nation where antiquities are found, even if foreign universities and institutions may be allowed to participate in such investigations. Officials of the Government of the Republic of Türkiye have been pro-active in guiding the explorers towards a scientific investigation under the oversight of Turkish universities. The Turkish museum authority has granted Ahmet Ertugrul a license to preserve the antiquities in a private museum in the scientific center just south of the mountain commonly referred to as Noah's Village. This museum will be essential for the archaeological investigation of the sites on the mountain.

Fig 29 - Artifacts in private museum just south of Mount Ararat

One thing holding back the granting of an archaeological license to this coveted site is that such investigations are expensive, none so much as a complex site near the top of a dangerous mountain that is more than three miles above sea level. Excavating the site requires new techniques and methodologies in the relatively new field of Alpine archaeology. It is important that this be done in the most careful manner because excavating any site changes the site, potentially loosing forever priceless archaeological data. How much more difficult a site that resides inside a glacier! The site must be stabilized not only to protect the ancient wooden structure, but also to protect and provide suitable working conditions for the archaeologists and scientists who work at the site. This will likely be the most expensive archaeological site ever investigated.

Almost all archaeological sites are privately funded, the job of raising the funding being the responsibility of the archaeologist granted the right to excavate a site. The scope of this project however is far beyond the capability of individual archaeologists. Historically, archaeological projects related to the Bible have obtained most of their funding from America. This site may be no different because one of the requirements for awarding an archaeological license is that sufficient funding be available before the Government grants any institution this coveted archaeological license.

The archaeological profession likewise grants the first right to interpret sites to the archaeologist granted the license to survey and excavate a site. Expert archaeologists of the same era and region occasionally challenge interpretations while independent prehistorians such as Great Britian's Graham Hancock and myself make wholesale challenges using new paradigms for understanding the past. I have of course challenged the interpretations of almost every prehistoric archaeological site in the world, however much I respect and value the expertise and discipline of this important craft upon which my work depends.

In order for archaeologists to determine whether this ancient wooden structure might be Noah's Ark, they must not only consult the Bible which is the most ancient text concerning Noah's Ark on Mount Ararat but they must also have a geological and archaeological understanding of Noah's Flood. That includes understanding the regional and worldwide migrations following the Flood. They will need this to develop a methodology for testing whether these remains might be Noah's Ark and that requires the new paradigm put forth in this book. For such purpose, I published *Ancient Light From This Mountain*[55] based on a monograph that I wrote for one of the archaeological conferences devoted to migrations associated with this mountain.

For these ends, I do not hesitate to give my preliminary interpretations of artifacts associated with the recent discovery which I believe that I can visibly identify as being of types defined by archaeologists of this region and period. Even so, a careful identification and thorough investigation must come through the official archaeological process to which we all look forward. As often the case, archaeologists may continue to differ concerning certain interpretations.

Confident as I am that this site is Noah's Ark, I am not infallible. Should there be found evidence that this discovery is not Noah's Ark, I will be the first to proclaim it. To this end, my method for determining whether this structure might be Noah's Ark is the same as was my methodology for investigating whether there existed the archaeological evidence of Noah's Flood. Thus, I aim to find various methods of showing why this ancient wooden structure cannot be Noah's Ark! Truth is more important to me than being right. But I have also discovered this to be an astonishingly productive way of discovering truth, why I choose to address rather than avoid the most difficult questions. It is the

---

[55] Philip Ernest Williams, *Ancient Light from this Mountain* (2022, CLSP)

reason that I choose to investigate the matter of the historicity of Noah's Flood which appeared the greatest scientific challenge to the Bible.

Even in cases where my particular observations in the following chapters may prove faulty, this does not necessarily impact the larger case of whether this ancient wooden structure is Noah's Ark. That would only point to limitations in my present understanding, especially owing to the current lack of a more comprehensive, detailed, and expert investigation. While we have much to learn about this ancient wooden structure and the origins and history of mankind, as I explain below I do believe that we have excellent reasons for believing that found the very remains of Noah's Ark.

47

A Rare Place Capable of Preserving an Ancient Wooden Ship and the Advanced Technology of the Antediluvian World

The site that Sheila Bishop and I visited is Site B, which rests inside a glacier of volcanic rock and ice. During summer thawing, the glacier moves this part of the ship sometimes causing it to surface and be visible in the region surrounding the mountain. That happened as recently as 2019. But in most years, all the sections of the ship are completely buried which not only protects the ship from the rocks which continue to tumble from the top of the mountain, but also from Ark searchers and those who would carry from the site the world's most valuable antiquities.

Fig 30 – 5,000 year old Otzi the Ice Man when found under a glacier in the Alps

Though the Ark has been damaged by the 1840 earthquake, by the moving glacier, and by waters flowing from the melting glaciers, its preservation is due to the frozen state of this ice-capped mountain which preserves wood even from the waters of the summer thaws of the glacier. We can compare its preservation to that of Otzi the Ice Man (Fig 30), who was found in the Italian Alps back in the nineties in similar conditions, as also the carcasses of mammoths found in Siberia (c.f. Fig 19, Chapter 18) all of which date to about this time. Though Otzi was likely a rustic hunter or warrior, his weapons and finely crafted clothing and sandals would be the envy of today's hunters and hikers.

Superficial observation of the site that I visited might cause one to suppose this ship was constructed from recently fabricated timbers as one might buy from Lowes or Home Depot. But a closer study of things like chisel marks on the wide boards and the square wood fasteners reveal a very different kind of fabrication than used today.

Fig 31 – State of the art antediluvian fabrication

Why many suppose the Ark belonged to an era of primitive technology is that many have suppose that Noah and his family lived in the Old Stone Age when technology is presumed to have been very primitive. However, the Bible does not reveal early man as being so primitive. Both Young and Old Earth Creation Scientists follow mainstream science's fictional narrative of cultural evolution! We should instead use the light of the Bible. According to the narrative of the Bible, cities are soon to appear in the history of man as do advanced artifacts. Consider the family of Cain, the first son of Adam who seems to have had many relatives at the time he killed his brother Abel.

... Cain was then building a city, and he named it after his son Enoch. To Enoch was born Irad, and Irad was the father of Mehujael, and Mehujael was the father of Methushael, and Methushael was the father of Lamech. Lamech married two women, one named Adah and the other Zillah. Adah gave birth to Jabal; he was

the father of those who live in tents and raise livestock. His brother's name was Jubal; he was the father of all who play stringed instruments and pipes. Zillah also had a son, Tubal-Cain, who forged all kinds of tools out of bronze and iron. Tubal-Cain's sister was Naamah. [Genesis 4:17-22]

As I have explained, the origins narrative of mainstream science was created from an imaginary development of technology based on the anthropology of living tribes and accordingly used to temporalize and vastly expand the antiquity of man through supposing a long Ice Age.

The remains of the Ark give us an unprecedented window into the advanced state of the antediluvian world, as we also see in the case of the Pyramids. We know that the stone ruins of ancient civilizations were built to remarkable precision. However, it is mostly the worked stone which survive at such sites. Far more can be understood about the antediluvian world by the rare preservation of organic artifacts as in the case of Otzi, but especially from the extensive organic remains of this ancient ship.

Though this ship has been visited many times even before the time of Christ as we can tell from artifacts brought to the site, its location near the top of a high and dangerous mountain has assisted its preservation. Far more have wished to visit this sacred site. That however required a means of travel and safe passage to the most remote part of Eastern Turkey which until recently has been one of the most dangerous parts of the world. The historical danger was due to the fact this mountain served as the boundaries of kingdoms and empires even before the time of the ancient Persians. Fortunately, the conditions are now such that the Government of Turkey can open this mountain to scientific research and tourism including the beginnings of human civilization.

Another factor leading to the preservation of this site is that regardless of the means or travel, mountains as high as Mount Ararat were rarely climbed until the last century. Mount Blanc, Western Europe's highest mountain, was not climbed until the late eighteenth century. Fortunately, from the time of modern travel in the nineteenth century, the Ark has been completely buried. Nor can the ancient wooden structure be seen from aircraft, drones, or satellite photos. Locations of the Ark supposed identified by such photos are nowhere close to the location of the ancient wooden structure! It is as if a higher power has been protecting this site until such time for it to be revealed to the world.

48

FOUND AT THE SPECIFIC RESTING PLACE OF NOAH'S ARK MENTIONED IN THE BIBLE

" and on the seventeenth day of the seventh month the ark came to rest on the mountains of Ararat. on the first day of the month, were the tops of the mountains seen." [Genesis 8:4-5]

Fig 32 – Viewing Little Ararat from Greater Ararat

I mentioned in the first chapter how the remains of the Ark were found at the one tiny spot on our large earth's surface matching the biblical account of the Ark's resting place. While this alone does not prove that these remains are those of Noah's Ark, it is nonetheless most remarkable, even more so due to the fact that this is the earliest specific geographical place mentioned in the Bible that can still be identified.

There have been numerous historical reports of the Ark remains still being found on this mountain, as those of the Babylonian historian Berossus dating from three centuries before the time of Christ. The first century Jewish historian Josephus mentions Berossus's report.

Some have rashly claimed that the ancient reports did not concern the mountain now called Mount Ararat, but rather some other mountain, possibly Mount Cudi near Turkey's southern border with Iraq. However, Josephus can only have referred to what is now Mount Ararat when he reports the Ark as 'resting on the highest parts of the mountains of Armenia.'[56]

The existence of the Ark still resting on the mountain which we now identify as Mount Ararat was part of the encyclopedic knowledge presumed by the finest historians through the Christian centuries, including Theophilus of Antioch (second century), Faustus of Byzantium who wrote the history of Armenia (fourth century), Epiphanius of Salamis (fourth century), and the great John Chrysostom (late fourth century). These reports continued through the Middle Ages including those of Isidore of Seville (sixth and seventh centuries), and Vincent of Beauvais (thirteenth century).

The Ark was also mentioned by traveling monks, merchants, and knights from the later Middle Ages including the Franciscan William of Rubruck (thirteenth century), Sir John Mandeville (fourteenth century), all the way to Marco Polo who describes the cup-shaped mountain upon which the Ark sits. The cup-shaped mountain can only have been Mount Ararat. The Silk Road to China begins just below Mount Ararat.[57] The Saint James Monastery on the north side of the mountain, long the guardian of the Ark, is of course also testimony to Mount Ararat being the site of Noah's Ark.

Most ancient and famous biblical sites were discovered from reports such as these, especially from traditions remembered by the native peoples living near the sites. However, it has only been from the time of Napoleon that Western archaeologists learned of such

---

[56] Josephus, William Whiston's translation, *Against Alpion,* Book 1, Chapter 19, passage 130

[57] The best documentation of these reports are published in the *Quest for Noah's Ark (*1972, Bethan Fellowship) by John Warwick Montgomery, colleague of Dr. Norman Geisler.

traditions and were able to identify biblical sites, beginning in Egypt and the Middle East. Even in the case of Palestine, biblical archaeology dates only from the nineteenth century. But this site is located in a part of the world of which most Western archaeologists are unfamiliar. Hence, the neglect of the place of origins of all the earth's people which is at last getting the attention it deserves. This region is on its way to becoming the most sought place in the world for archaeological tourists.

So far as access to biblical sites, it seems that the first site mentioned in the Bible shall be the last to become known to the world. The good news is that while peace is disappearing in much of the Middle East, for the first time in history, peace is flourishing in the region of this mountain. This is not only good news for the people near this mountain but also for the world!

49

IT DATES TO THE ERA OF THE FLOOD!

Using the genealogies given in the Bible, Archbishop James Ussher dated the Flood to 2349BC. Ussher dating was based on the Masoretic text and also by reckoning the creation of Adam as being exactly 4000 years from the birth of Christ. Chronologies using the Greek texts push back the Flood to centuries earlier. But whatever text is used, several assumptions have to be made to calculate the exact date of the Flood from the Bible. Based on radiocarbon dates of certain Flood burials as I mentioned in the *Archaeological Evidence*, the Flood dates almost exactly to the 2349BC date calculated by Archbishop Ussher.[58]

The first and largest board from this ancient wooden structure sent to an Iranian dating laboratory yielded a calibrated radiocarbon date of about 4800 bp. The larger timbers used for building the Ark would have been growing for hundred years before they were harvested. Thus, the date of this timber exactly matches the time of the Flood. Smaller pieces of wood subsequently sent to the same laboratories date from after the time of Christ. Thus, a more comprehensive radiocarbon dating of this ancient wooden structure

Fig 33 - Stone vessel of Early Bronze Age

---

[58] Williams, Philip, *The Archaeological Evidence of Noah's Flood* (2011, CLSP), pp 273-275

waits for the Government to issue an archaeological license.

While we look forward to more extensive radiocarbon dating of this site conducted under expert oversight and authority, I discussed in Chapter 4 some of many problems associated with radiocarbon dating. This explains why I have far more confidence in artifact dating. The black stone bowl of Fig 33 (previous page) was found inside Site A. Such bowls as made from hard basalt were common in the Early Bronze Age but disappear in the Middle Bronze Age that immediately follows the Flood.

Fig 34 – Rim-handle jar found in Ark

Early Bronze Age sites are the most advanced of the antediluvian world. The antediluvians could precisely cut, transport, and lift even very large stones as those in the Pyramids by means that we have yet to understand. Likely, a lot of slavery was required for such industries, which would explain why Noah's family did not preserve this technology.

Pottery similar to the rim handled vessel of Fig 34 that was found in the structure are also found at Anatolian sites belonging to migrations immediately following the Flood. The handles of these vessels allowed them to be strapped to donkeys and horses.

Fig 35 – Pottery shard impressed with 'tree of life' motif

An even more interesting ceramics found inside this structure is that of the shard of Fig 35 upon which is impressed with the 'tree of life' motif. This 'tree of life' motif is also found on cylinder seals and pottery from the Sumerian and Akkadian eras which date from just after the Flood. While the point of discussion here concerns the dating of this site to the era of the Flood, this and similar examples found in Sumerian and Akkadian sites show how the account of events from the Garden of Eden are being passed down to the survivors of the Flood all of whom descend from Noah. Likely, the tree of life inspired the epic of Gilgamesh, who sought out the survivor of the Flood, who we may suppose is the same as the Noah recorded in the Bible.

I mentioned in Chapter 7, the bevel-rim bowls associated with Late Uruk which appears to be the site and era of the Tower of Babel. These mass-produced bowls were made from a wooden mold, though archaeologists have never found one of the wooden molds. This is because wooden artifacts do not last for thousands of years. However in Fig 36 below, we might be looking at some of these wooden molds together with containers which held the wet clay to be used for these molds. According to Parasut (who you see in this photograph) to whom I showed examples of bevel-rim bowls, these are among the ceramics residing inside the Ark. What powerful evidence that in this photo, we are looking inside Noah's Ark!

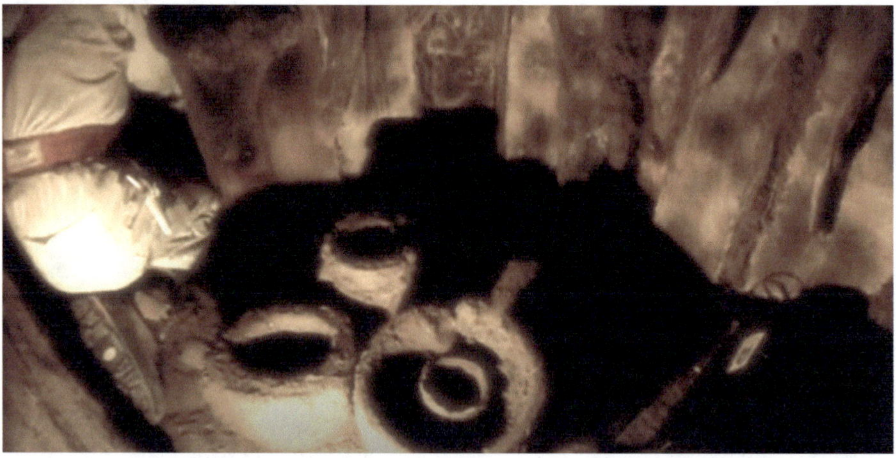

Fig 36 - Wooden molds for making bevel-rim bowls?

More evidence that this ancient wooden structure dates from the time of the Flood are the presence of Neolithic chickpeas as I discuss in Chapter 53 below. It seems that we do have a ship that dates almost exactly to the time of Noah's Flood and contains artifacts of the very material culture from which began the origin of nations!

50

EVIDENCE OF THE ANTEDILUVIAN CLIMATE

Dendrochronologists use the variable patterns of tree rings to correlate the counting of tree rings from trees of different ages that grew in the same forest, thereby hoping to date wood preserved from ancient times. These patterns are due to our variable seasons and rainfall. In years of greater and more extended days of rainfall, the growth rings are wider but are narrow in years of drought.

The earth's seasons are due to the earth's axis tilting towards the sun in summer and away from the sun in winter. Those living in the northern hemisphere experience summers and winters exactly opposite of those living south of the equator. Our variable seasons are also the cause of rainfall as moist air from the tropics is carried north in the summer. As cold air sweeps down from the poles, this moist air condensates to become rain.

In Chapter 57 below, I will show how the evidence of the antediluvian world points to a different kind of hydrological cycle before

Fig 37 – Straight even growth rings on post inside structure

the Flood, as mentioned in the second chapter of Genesis. Prior to the Flood, the earth was watered by springs from great reservoirs just below base rock near the surface of the earth. At that time, there was not enough difference in temperature between summers and winters to cause rainfall. At the time of the Flood, most of the water in these reservoirs was suddenly emptied due to a sudden tilt of the earth's axis which resulted in our present summers and winters. Those who aim to find a cause for the Ice Age also attribute it to a change in the earth's axis though one that is smaller and supposed as occurring in a regular fashion.

An antediluvian climate with invariable seasons may explain a remarkable observation by a Dutch dendrochronologist who pointed out to Gerrit Aalten the evenly space growth rings as can be observed in the photographs in Fig 37 above and Fig 38 below of boards in this ancient wooden structure. Such evenly spaced growth rings was something that the dendrochronologist had never seen.

One may suppose this should have been observed in the trees older than about five thousand years that are used for calibrating radiocarbon dating because evenly spaced growth rings make correlations difficult. However, counting tree rings in the world's oldest trees is hardly the straightforward process that is widely supposed. It may be time to revisit the assumptions and methodologies with a closer examination of the tree rings from most ancient times and the way we count them. The timbers preserved in this ancient wooden structure will be useful for this purpose.

Fig 38 – My photograph of evenly spaced growth rings

51

The Ancient Wooden Structure is Not Only a Ship but is Built Like Noah's Ark!

So make yourself an ark of cypress wood; make rooms in it and ***coat it with pitch inside and out****. This is how you are to build it: The ark is to be three hundred cubits long, fifty cubits wide and thirty cubits high. Make a roof for it, leaving below the roof an opening one cubit high all around. Put a door in the side of the ark and make lower, middle and upper decks. [Genesis 7:14-16]*

The Hebrew word *teba* (or *teva*) translated here as *ark* is also used for the little boat in which baby Moses floated on the Nile. This Hebrew word seems to have been transliterated from some language ancestral to biblical Hebrew. Likely, the meaning of this word in the ancestral language was *boat,* or *ship*.

Fig 39 - The interior of site I visited when first opened in the Fall of 2011

Three different sets of observations point to this ancient wooden structure as being a ship. The first is that the outer walls are bowed slightly inward as shown in Figures 40 and 42 below. The bow indicates that these walls are the hull of a ship. Some have supposed that the Ark was a rectangular barge, but a rectangular barge does not have the stability of a bowed ship. For a ship this size, building the hull with long and wide horizontal boards is a more reliable and practical design than bending narrow vertical boards which might not want to stay bended for the entire voyage.

Fig 40 – The ship's black bowed hull extending into the darkness

I presume these horizontal boards are not only secured to the great timbers that form the bulkheads but are fastened to one another using the same long square pegs as fasten internal boards that I did see and touch (Fig 31). At the time of my visit, these boards were coated with ice, as shown in Figs 42 and 43 below. This coating of ice had built up on the wood inside the ship due to moist air coming from the passage that had been opened by the climbers. Fortunately, the climbers photographed the same location when this site was first opened in 2011 as shown in Figures 39 and 40 above. Keep in mind that this is but one floor of the three sections of the buried ship. I could only view a portion of this section, which isn't the largest. Even so, in Fig 40 we see the hull extending far into the darkness.

The second set of observations is the black coating on the hull which is clearly visible in these early photographs and points to this ship as possibly being the pitch-covered Ark of Noah. There has been unnecessary debate as to just what was meant by 'coat it with pitch inside and out.' The pitch was needed for sealing the hull of the Ark.

Fig 41 – Wooden chest secured to shelf

While the pitch on the external walls would have blocked water from soaking the boards and entering whatever small cracks existed in the wall, the internal pitch would have helped seal the Ark from the penetration of moisture and block whatever moisture did manage to penetrate the exterior hull. While we observed some of the pitch as having dropped on internal boards, pitching the internal walls would have caused unnecessary problems with the clothing worn by Noah's family as well as the fur and feathers of animals and birds.

Fig 42 - Author's photo of ice-covered hull

Further evidence that this wooden structure is a ship is the fact that the contents are secured to the structure itself, as the case of hole in the shelf which secures the black stone vessel in Fig 33 above and also the chest secured to the shelf as shown in Fig 41.

The ancient wooden structure is broken into three sections which are buried below the surface of the mountain. Because I entered the structure through a tunnel dug through the rocks, I could not view the external walls. However, I had an excellent view of an internal wall of this structure.

Fig 43 – Viewing lower deck from Ark's second deck

A cubit is the distance from the elbow to the tip of the finger, often supposed to be about 18 inches. But we cannot be sure of the exact measure from such ancient times. Based on a cubit of 18 inches, the Ark would have been 450 feet long, 75 feet wide, and 45 feet high. (137 meters by 23 meters and 14 meters high). Due to having been broken and not fully explored, the exact construction and dimensions of the broken ship have yet to be measured. That will have to wait for an official archaeological excavation by the team granted the privilege to survey and excavate this ancient wooden structure. But according to my rough floor-to-ceiling guestimate, the Ark is somewhat larger than our 18-inch cubit and perhaps the size of Ken Ham's Ark Encounter in Williamston, Kentucky.

During my visit on November 9, 2014, the date stamped on many of these photographs, I was only able to explore the level passageway which extended from the bottom of our entrance tunnel, the bowed black hull serving as the right wall of this passage. The slanted wall immediately to our left seems to have been formed when the upper deck collapsed due to the many tons of rock that had fallen onto the Ark. I was however able to peer into a lower deck from the hole formed from this fallen wall (Fig 43). I thus surmised that this section of ship had upper, middle, and lower decks, thus built according to the instructions which the Lord gave Noah for building the Ark.

52

EVIDENCE OF HAVING BEEN A FLOATING ZOO

Take with you ***seven pairs of every kind of clean animal,*** *a male and its mate, and* ***one pair of every kind of unclean animal,*** *a male and its mate and also* ***seven pairs of every kind of bird,*** *male and female, to keep their various kinds alive throughout the earth [Genesis 7:2-3]*

Fig 44 – Example of boards with 7 pegs

Some of the most convincing evidence that points to this structure being Noah's Ark is finding unexpected things which point to such a conclusion. For example, we typically think of a single pair of each kind of animal being saved by the Ark. However, as the above verses show, the Lord also instructed Noah to take aboard the Ark seven pairs of each kind of clean animal. This might explain the seven pegs fastened to beams that extend across some of the rooms in this structure (Fig 44.)

We might suppose that these were used to tie pairs of clean animals in the open spaces of the Ark whether to keep them from crowding and suffocating, or for tying them while drinking water as I explain in the next chapter. Among the clean animals are sheep, goats, horses, donkeys, and cattle.

Fig 45 - Displaced post with slots for cage slats

The dislodged wooden post that I observed on the floor shown in Fig 45 above has slots perfectly spaced for cage slats. One wonders just what kind of animals were held in the cage supported by this post. The close spacing of these slats suggests that it may have been one of the larger wild animals.

Perhaps the best evidence of this ancient wooden structure having been a floating zoo was an acute odor coming from a room when the explorers opened a certain door. It seems this permanently frozen site preserved urine and animal droppings as fresh as when they were

Fig 46 - Exploring the interior rooms of the Ark

dropped. The chemical emissions from these droppings concentrate inside the enclosed compartment. After opening the door to this compartment, the explorers were so overcome that they had to be hospitalized for an entire week. This explains the masks worn by the explorers in some photos from the site.

53

FITTED WITH THE RIGHT SUPPLIES FOR A FLOATING ZOO

Two of every kind of bird, of every kind of animal and of every kind of creature that moves along the ***ground will come to you to be kept alive.*** *You are to take* ***every kind of food that is to be eaten and store it away as food for you and for them.****" [Genesis 7:20-21]*

Fig 47 – Neolithic chickpeas scattered on the floor

Note the animals came to the Ark on their own to be saved! The white pellets you see in Fig 47 below are neolithic chickpeas. These are smaller than today's chickpeas but even more nutritious. The interesting thing about chickpeas is not only are they an excellent source of protein, but they can be eaten by almost any animal.

Fig 48 – Basket found in Ark

While the descendants of these animals aboard the Ark will develop specialized diets based on their eventual habitats, this was not the case of their genetic ancestors. Keep in mind that the head zookeeper of this floating zoo is the same as created these animals.

Fig 49 – Large room in bow of ship used for food storage.

While aboard the Ark, the animals did not have to hunt or forage for their food which will be brought to their cages and stables. The basket shown in Fig 48 that is woven from coarse strands is filled with what appears some nutritious cane. Likely, the scoop sitting on the ledge of the large room of Fig 49 was used for scooping grains and feeding the animals. Consider how much food may have been stored in this one large room.

Fig 50 – Some of the Arks watering

While continuously resting, animals do not require a lot of water. However they do need to occasionally drink in order to maintain their metabolism. This explains the stone bowls in Fig 50. This suggests that the Ark had water tanks. I suppose they were built into the walls along the water-proof hull of each deck. The seven pegs may have been used to tie the animals after being led to these bowls. However they operated, Noah and his family were well occupied carrying food and water to the animals, while scooping their waste and depositing it in that room from which the strong emissions came.

Though there remained these traces of the food supply that Noah and his family brought aboard this ship, it was not without reason that the Lord guided this Ark to rest on the peak of such a high mountain as Mount Ararat. This gave several extra months for vegetation to grow in the fresh and fertile soil that covered the mountain immediately after the Flood. In the years immediately following the Flood, the climate on the top of the mountain was much warmer. This is because the ocean levels were also much higher before the ocean basins slowly stretched and sank as I will explain in Chapter 60.

Noah and his family would have released the animals in appropriate order. The first to be released were the birds, herbivores, and marsupials. Months later, the carnivores would have been released. To this day, the animals are still being guided and fed by God and protected by the angels since reaching their destinations. Apart from God's supernatural guidance and power on behalf of Noah's family and the animals, none of this could have happened. Thus, we have found at this site near the top of Mount Ararat manifest evidence of one of the greatest miracles recorded in the Bible.

54

Impossible for this Ship to Have Been Built on the Mountain

One of the reasons for rejecting the possibility that this ancient wooden structure was built on the mountain is that it is found in a wrecked and fragmented state at 3 widely spaced locations. Supposing that it would or even could have been built inside a dangerous glacier in its presently wrecked state should not be a serious consideration. Just consider the difficulties and cost of such an undertaking. And for what purpose! After visiting the site, Sheila Bishop remarked, 'It insults one's intelligence even to suggest that this large structure could have been built inside this glacier.' As you may have heard, such claims were made by rival Ark searchers whose plans were disrupted by the announcement of this discovery. However, as Sheila and I know, it is dangerous just to visit this structure.

If indeed built on the mountain, the ship had to be built on or under the glacier on top of the mountain, then somehow pushed down the mountain to create the current wreckage. Some archaeologists do consider the possibility that this Ark was built as a pilgrimage site on this mountain where the Bible tells us that Noah's Ark came to rest.

Building a ship almost one half the size of the Titanic on level ground would be no small undertaking. Fabricating such a large ship more than 5,000 meters (16,000 feet) above sea level which is far above the tree line where no vegetation grows, would have been a truly monumental, indeed impossible feat. Simply transporting the large timbers that we observe in this ancient wooden structure would have required building a road for whatever carts and work animals might be able to ascend to such heights. The special breed of horses that can climb these heights reach their limit at about 4,000 meters (14,000 feet).

Had the construction taken place in the last few millennia, the road and construction site would have to be built through the glacier that covers the top of the mountain. Not only would the workers need to clear the glacier, they would have to clear it again and again after each snow. Near the top of this mountain, snows occur every month of the year. The workers would also need a work camp near the site of construction. But to this day, there exists no road to the top of the mountain. The construction materials for such a project would have been preserved under the same frozen conditions as the Ark. So far as we know, no such materials have been found. The pilgrimage site would have surely been built from the era of the oldest artifacts found in this structure which date from before the Flood. Surely, Noah could have found a more convenient site for building the Ark!

Fig 51 - I took this photo inside this underground "cathedral"

As mentioned in Chapter 48, there are indeed reports dating from before the time of Jesus of pilgrims visiting this site. Some artifacts found in the structure date from a later time. If such an undertaking did take place, it would surely date to when the great Byzantine Emperor ruled this region. Those who undertook such a job would also have been archaeologists with collections of artifacts dating from the time of the Flood. Though ancient kings and emperors were never shy about documenting and crowing over their construction projects, there are neither monuments nor historical reports concerning building the Ark on this high mountain.

If such a ship could hardly be fabricated on the mountain, neither could it have been transported to the top of the mountain. The only remaining explanation for this ship being found on the top of this high mountain is that it floated up there.

55

IT IS THE EXACT PLACE OF ORIGIN OF THE WORLD'S HISTORICAL NATIONS

However well this recent discovery on Mount Ararat conforms to the exact resting place, the exact size, design, and time of the Flood as can be calculated in the Bible, the only thing which can prove the ship that I visited is Noah's Ark is discovering that that all the earth's historical nations and land animals can be traced from this ancient wooden structure.

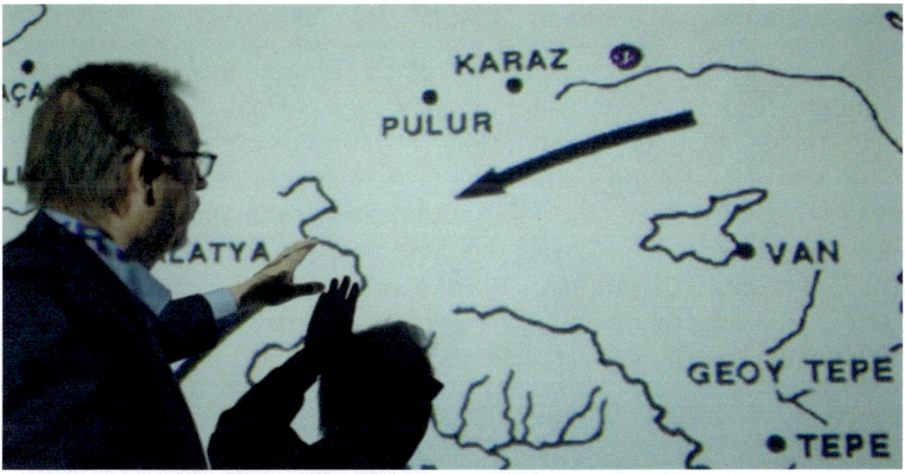

Fig 52 – Migrations from Mount Ararat

While we must wait for the studies concerning all ancestral species being traced from this mountain and region, we already know this is the land of origins of numerous domesticates, including grapes, apricots, wheat, barley, sheep, goats, and horses. Keep in mind, this mountain is just north of the apex of the Fertile Crescent where began agriculture and civilization.

The specific set of ceramics that belonged to the migrations from this mountain that I discussed in Chapter 7 included red-black burnished pottery, which obtained its coloration due to being baked in open fire. Animal figurines are also found among the ceramics from these sites as if Noah's family were sharing with their children what they had seen aboard the Ark. When I heard the announcement of this discovery, I thought how marvelous if some of these ceramics might be among the artifacts in the Ark! At the time. I did not know whether any artifacts had been found inside the structure. This was but a fond hope. I was thus thrilled to learn from the first photos of ceramics that Parasut shared included the shard of animal figurine shown in Fig 53 conform to the family of ceramics that I had traced from this mountain. This of course explains my interest in the discovery from the time it was announced and also my confidence that Noah's Ark has been found. The migrations from the area of this mountain from which the historical nations all across the earth derive are sufficient to prove that the Ark did in fact rest some place on this mountain.

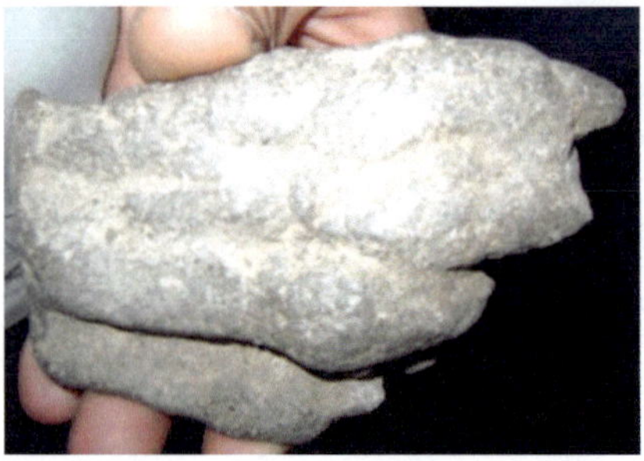

Fig 53 – Figurine found in Ark

How likely might it be that the Ark rested someplace else in the mountains of Ararat? Much attention has been given to the Durupinar site that is just south of Mount Ararat. The contours of this site resemble a large ship, if longer and wider than the biblical dimensions. The pointed shape of this site has been supposed as reflecting the reed ships of southern Iraq and giving more credibility to the shape of the Ark given in the Epic of Gilgamesh! This site was promoted by the late Ron Wyatt of Knoxville, Tennessee and has continued to be exploited as the site of Noah's Ark since Wyatt's death.

I have often visited this site which is a few miles east and south of Noah's Village. The ridges forming the contour of this site resemble similar formations along the same road as the Durupinar site. This is because all these formations were the result of the same 1948 earthquake! Nor are the tall standing stones with drilled round holes near the top that have been supposed as drogue stone anchors for Noah's Ark any nearer the Durupinar site than they are to the Ark that I visited. However one chooses to explain the Durupinar site, one must look up from this site to see the tops of the mountains of Ararat. This means that the Durupinar site would have been deeply covered by water when Noah’s Ark landed on Greater Mount Ararat, the highest mountain in all Ararat.

I once considered whether the Durupinar site might have been the dry dock where Noah built the Ark. But there is no evidence here of the kind of advanced construction that was used for the Ark that I visited. Based on the Kura-Araxes artifacts and radiocarbon dating, it seems that the Ark was built someplace in what is now the Republic of Georgia, which is between the Black and Caspian Seas. How convenient for the Ark to have been located at a place that is in the central part of the three Old World continents and continues to be a place of the most diverse and varied habitats.

It seems that immediately after the Ark landed, Noah and his family returned to their home in present day Georgia, then part of what the Sumerians called the land of *Aratta* from which place Nimrod led settlers down the Tigris and Euphrates Rivers to the land of Shinar, which the Greeks would later call Mesopotamia. These settlers would have needed an expert hunter to explore and lead them to new lands in a time when dangerous wild animals then only recently released from the Ark were still roaming these lands. Houses built during this era enclose a large courtyard, which are convenient for protecting people from dangerous animals, though not so much from hostile enemies which had only recently been entirely removed from the earth by the Flood. There is no other way to explain the evidence from the Ancient Near East aside from the light of Noah's Ark and Flood exactly as put forth in the Bible.

Part VI

The Waters of Creation and the Flood

56

A Geology and Cosmology Based on the Waters of Creation

In the beginning God created the heavens and the earth. Now the earth was formless and empty, darkness was over the surface of the deep, and the Spirit of God was hovering over the waters. [Genesis 1:1-2]

Ever since the Copernican Revolution, at such time the earth was determined to be a planet in a much larger solar system, the waters of Creation mentioned in the above creation account have been read phenomenologically rather than literally and with the waters of the deep supposed as belonging entirely to the earth. This goes back at least to Isaac Newton who explained the Creation account in Genesis as being that of an observer standing on the surface of the earth. As have most since this time, Newton was interpreting the creation account from a superficial observations of the heavens which fails to reveal the great mass of water which God used to form both the heavens and the earth.

The account however is clearly that of the Creator who revealed this account to his prophets. If the aim of this account is to explain the formation of the earth and the creatures that live upon the earth, the perspective of the Creator could hardly be that of his creatures. We should thus remember the many passages of Scripture that refer to the Lord as spreading out the physical heavens. Hence, the Creation account is better read as teaching the heavens as also having been created from water by water with space being created by the separation of the waters on the second day of Creation. This is in line with the words of the Apostle Peter who personally knew the Creator.

"... long ago by God's word the heavens came into being and the earth was formed out of water and by water. By these waters also the world of that time was deluged and destroyed." [2 Peter 3:5-6]

Such neglect of the universe as having being created from water led to what I was taught from my earliest days in science: that water is rare in the universe and in limited supply even on the earth. Recent observations are revealing that nothing can be further from the truth. But this myth has created what seemed a problem for Noah's Flood. Where is water that could have covered the earth's highest mountains? And after the Flood, where did all that water go?

THE CURRENT SCIENTIFIC NARRATIVE OF THE WORLD'S BEGINNINGS

The scientific narrative of the origins of the universe commonly referred to as the 'Big Bang' explains the primary material of the universe as being hydrogen atoms that were created from the energy of the Big Bang as soon as the universe cooled sufficiently for atoms to exist. Supposed as supporting this is the fact that the light emitted from most of the visible universe is from hydrogen atoms. The second most abundant observed material is helium which is formed from the fusion of hydrogen. Far less of the mass of the universe is supposed as consisting of heavier atoms which are believed to have been formed in the interior of stars, with the heaviest elements coming from the collisions and explosions of stars.

The earth, on the other hand, is believed to have been formed by the gravitational coalescence of melting meteorites composed of heavy metals such as iron, nickel, and lead. Thus, as currently taught, the earth's core is composed of these heavy metals in liquid and solid states. The earth's mantel and solid crust is supposed as formed from the oxides of lighter atoms such as silicon, calcium, and aluminum. But there is no scientific consensus concerning the source of the earth's water, whether it came from the heavens or from inside the earth.

A problem with the current theory of the formation of the earth is that meteorites lack the mass to have been formed from star dust. Thus, they require terrestrial bodies for their formation. The meteorites that impact the earth come from asteroids or else have been blasted into space by large comets or meteorites which create those craters on the surface of terrestrial bodies such as the moon and Mars. Terrestrial bodies are thus needed for creating meteorites. But how were terrestrial bodies formed before there were terrestrial bodies from which meteorites could be blasted into space? Again, note that all this speculation about the material composition of the heavens and the earth are from superficial observation of the earth and visible universe.

FINDING WATER DEEP INSIDE THE EARTH

I will summarize here what led me to discover the source and destinations of waters sufficient to cover the earth's highest mountains. It began with a question that occurred to me as soon as I had completed my investigation of the archaeological evidence of the Flood. Knowing that rocks are heavier than water, I wondered why continents float higher upon the earth's mantle than do the oceans. According to the principle of isostasy, this should not be the case. Rather, it suggested that the continents contain more water than do the oceans.

In 2011, I published my prediction that more water is to be found beneath the earth's continents than in all the world's oceans.[59] In 2013, we learned that more water exists under North America alone than in all the earth's oceans! Most of this water exists in crystalline form. However, being lighter than rock, how did water get so far below the surface of the earth so as to form these crystals? As I noted in Chapter 13 concerning the deepest hole ever drilled into the earth's crust, increasing amounts of liquid water were found in the lower parts of the Kola Superdeep Borehole. This science is in a state of rapid change but the simplest explanation for the internal composition of the earth is that of the Galilean fisherman. The earth was made out of water from water!

As I learned from one of the world's leading material scientists who hoped to recruit me for this science, we do not even know the properties of materials under such great pressures and temperatures as exist deep inside the earth. Only sixty miles beneath the surface of the earth, pressures are sufficient to create diamonds. At even greater pressures and temperatures as exist far below the earth's crust, water turns into a plasma state which is far more compressible than solid or liquid metals and rocks. Water existing in a plasma state deep within the earth will thus be heavier than metal or rock! Likely, the deeper and greater portion of the earth's interior is water as we have been recently discovering.

A WATER BASED COSMOLOGY

By the word of the Lord the heavens were made,
their starry host by the breath of his mouth.
He gathers the waters ... he puts the deep into storehouses. [Psalm 33: 6-7]

---

[59] Williams, Philip Ernest, *The Archaeological Evidence of Noah's Flood* (2011, CLSP), pp 264-265

If we have difficulty knowing what is deep inside our earth, how much less do we know what is deep inside the planets, moons, and stars? If however we trust the biblical account of origins, the interior of the planets, moons, and stars may also be water in a plasma state. Dark matter, which we know only from its gravitational effects, may however be ordinary water dispersed in space whether in a vapor, liquid, or frozen state as we know to be the case of comets. Just as the Scriptures tell us, the world's creation was out of water from water.

From a scientific perspective a water-based cosmology as I propose is 'just a theory.' So are the current theories of the origins of the earth and the universe which have many problems, have often changed, and will continue to change. None of the current theories is based on the authority of the revelation of the Creator. We should at least consider a literal reading of the account of Creation in the Scriptures to determine whether it might not only be a superior theory but a true understanding of the creation of the universe. Such a transforming theory is better classed as a new paradigm and judged by its specific, testable, and useful predictions as well as its simplicity and elegance on the basis of how well it integrates what is currently fragmented knowledge. On this basis, a water-based cosmology vanquishes the current cosmological theories.

A UNIFIED COSMOLOGY BASED ON WATER

Though water consists of bonded hydrogen and oxygen atoms, God can create things in a fully formed state as the case of the loaves and fishes by which Jesus fed the multitude. Thus, hydrogen and oxygen atoms would have probably come from the breakdown of water. Should the waters of creation be the same as we know on our earth, it would have trace minerals which we may suppose as having been formed from the initial burst of energy mentioned in the Scriptures, the light created on the first day of Creation.

A cosmology based on water can explain the size and nature of all the heavenly bodies in a simple and unified way as God separated the waters, then gathered water in various sizes to create the heavenly bodies. In the case of terrestrial planets and moons, the trace minerals are sufficient for the spherical body of water in a plasma state to be capped by an impermeable terrestrial crust formed from the trace minerals. The earth's oceans formed from the water that escaped above the terrestrial crust. Although the surface of terrestrial planets and moons show evidence of having once had flowing water, only our privileged planet has been able to retain its surface water.

Based simply on the size of cosmic bodies, the gravitational coalescence of water containing the same proportions of trace minerals as water on our earth, according to increasing size, will

1. form comets from the smallest bodies of gathered water
2. form terrestrial moons and planets from larger bodies of gathered water
3. form gaseous planets from even larger bodies of water
4. and form stars in the case of the largest bodies where the gravitational pressures are sufficient to ignite nuclear fusion.

Because of insufficient mass, smaller bodies of water become comets, though soon loosing most of their water into space. We already know that comets are mostly ice which melts as comets approach the sun. Larger bodies of water will obtain an impermeable terrestrial crust from the trace minerals, thus capping the water to become terrestrial moons or planets. Likely, the asteroid belt was initially a small terrestrial body that broke up releasing its plasmatic water while leaving its broken terrestrial crust as asteroids.

Due to the fact that the terrestrial cap will decrease in depth with size, larger bodies of water will have insufficient trace minerals to obtain an impermeable terrestrial cap and thus become giant gaseous planets. Still larger bodies of water will have sufficient compression for nuclear fusion to occur, thus becoming stars. Taking the biblical account of Creation seriously, we have a simple and unified cosmology capable of numerous practical predictions, including the 4 explained above.

Cosmology in the Light of the Miracles of the Six Days of Creation

Observed microwave radiation coming from all points of distant space is reckoned as confirming what is supposed as the world's beginning in what is called the *Big Bang*. However, the biblical account of this light as having been preceded by the creation of water does not require the breakdown of the laws of physics at the beginning of creation which in the case of the current scientific narratives requires the *ad hoc* theory known as *Inflation*. The trace minerals of the waters of Creation as mentioned above would have been created by this sudden burst of energy which has since shifted to the infrared spectrum. The separation of the waters on the second day of Creation created space. This space-creating separation of waters results in a red shift in the light from stars and galaxies as God's 'dark energy' spreads the heavens, accelerating in ways that no physical theory can explain.

The earth itself was formed from the waters below as the waters above created the space where the galaxies and stars would be created on the fourth creation day. The waters formed the space of the heavens even before the existence of stars on the fourth day! This earth-focused creation explains the so-called anthropological theory, why the earth is such a privileged planet. The earth is indeed a privileged planet, the focus of God's physical Creation. This explains why the Milky Way may be the oldest galaxy according to the measured age of some of its stars.

However poetic may seem the following account of Creation, it is also a scientific account according to a literal reading of the Scriptures:

First Day of Creation

In the beginning God created the heavens and the earth. Now the earth was formless and empty, darkness was over the surface of the deep, and the Spirit of God was hovering over the waters. [Genesis 1:1-2]

Though we are accustomed to imagining this as a creation in space, space has yet to be created from this formless and empty water that exists only in God's Spirit. There is no space because there has not yet been a separation of the waters of Creation. God's Spirit, which is his unseen presence in the physical creation, hovers over the waters from which God will create the heavens and the earth.

And God said, "Let there be light," and there was light. God saw that the light was good, and he separated the light from the darkness. God called the light "day," and the darkness he called "night." And there was evening, and there was morning—the first day. [Genesis 1:3-5]

Because God's Spirit was hovering uniformly over the water, this first day light was initially spread uniformly through the water. It was the separation of light from darkness that created a great burst of energy as the gathered light became intensely bright. The separation also created the darkness in the part of the waters from where the light had been removed. The background infrared radiation that we observe today came from this intense burst of energy as did the heavier atoms of the universe, the latter becoming trace minerals distributed through the water.

Second Day of Creation

And God said, "Let there be a vault between the waters to separate water from water." So God made the vault and separated the water under the vault from the water above it. And it was so. God called the vault "sky." And there was evening, and there was morning—the second day. [Gen 1:6-8]

God did not view the world as may have some Ancient Near Eastern peoples and pretentious biblical scholars. The word *vault* is better translated as *expanse,* as this verse points to the of their water into space. We already know that comets are mostly ice which melts as comets approach the sun. Larger bodies of water will obtain creation of space, as also in this verse from Isaiah:

He stretches out the heavens like a canopy,
and spreads them out like a tent to live in. [Isaiah 40:22]

While God transcends space, his Spirit is immanent in his Creation in such great measure as to extend across the physical universe. Gravity is the result of the holding together of the universe by its omnipresent Creator.

The waters below will be used to form the earth on the third day of Creation.

Third Day of Creation

And God said, "Let the water under the sky be gathered to one place, and let dry ground appear." And it was so. God called the dry ground "land," and the gathered waters he called "seas." And God saw that it was good. [Genesis 1:9-10]

Here begins the creation of the earth's continents as the trace minerals in the waters form the earth's terrestrial crust. The topmost crust is formed of basalt as the waters above the ocean basin cover the entire earth. The continents form as lighter granite, which crystalizes far below the earth's surface, rises through the heavier oceanic basalt and then above the oceans. The formation of the continents began at the earth's north and south poles due to being places of minimal centrifugal forces. As these initial continents increased in size, centrifugal forces from the spinning earth broke and spread these initial continents until they appear as we observed them today. In Chapters 59 and 60, I will show how the Flood better explains geological phenomena now reckoned as due to supposed plate tectonics.

Then God said, "Let the land produce vegetation: seed-bearing plants and trees on the land that bear fruit with seed in it, according to their various kinds." And it was so. The land produced vegetation: plants bearing seed according to their kinds and trees bearing fruit with seed in it according to their kinds. And God saw that it was good. And there was evening, and there was morning—the third day. . [Genesis 1:11-13]

In the early days of the universe, what is now low intensity infrared background radiation would have been warmer and closer to the visible spectrum of light. We may presume this light was sufficient for the growth of plants on the earth on the third day of Creation after land began slowly rising above the oceans creating the continents. Interestingly, chlorophyl has the ability to absorb light in a portion of the infrared as well as in the visible spectrum. After the topsoil from these decaying plants washed into the rivers and oceans, these plants turned into oil and gas. Thus, did plants thrive on earth prior to the creation of the sun, moon, and stars on the fourth day of Creation.

Fourth Day of Creation

And God said, "Let there be lights in the vault of the sky to separate the day from the night and let them serve as signs to mark sacred times, and days and years, and let them be lights in the vault of the sky to give light on the earth." And it was so. God made two great lights—the greater light to govern the day and the lesser light to govern the night. He also made the stars. God set them in the vault of the sky to give light on the earth, to govern the day and the night, and to separate light from darkness. And God saw that it was good. And there was evening, and there was morning—the fourth day. [Gen 1:14-19]

An additional miracle on the fourth day of Creation was required to gather the uniformly expanding waters and turn them into stars inside expanses of water which in turn formed the galaxies. Such creation of stars and galaxies in uniformly expanding space is something for which the current scientific narrative has no good answer! Keep in mind that the waters from which these heavenly bodies were made were spread from the Second Day of Creation. To create the sun, moon, and stars, various amounts of water had simply to be gathered in whatever part of the universe that God chose. After such gathering, gravity alone would be sufficient to form these bodies as explained above.

The universe is still expanding. Hence the red shift will occur in whatever region of the universe the stars are created. We might presume that the vegetation of the earth adapted during the fourth day of Creation for the animals that will be created on the fifth and sixth days.

Fifth Day of Creation

And God said, "Let the water teem with living creatures, and let birds fly above the earth across the vault of the sky." So God created the great creatures of the sea and every living thing with which the water teems and that moves about in it, according to their kinds, and every winged bird according to its kind. And God saw that it was good. God blessed them and said, "Be fruitful and increase in number and fill the water in the seas, and let the birds increase on the earth.' And there was evening, and there was morning—the fifth day. [Gen 1:20-23]

We see this fifth day Creation beginning with what is called the Cambrian Explosion, the sudden appearance of mollusks and shellfish, followed by the creation of vertebrate fish. The Cambrian Explosion is now believed to have begun about 540 million years ago. This is followed a few hundred million years later by the creation of the dinosaurs, which we now understand as the earliest birds. These changes in the size and appearance of birds came about through adaptations (but not through evolution!) of the same fully created species of dinosaurs. The Lord God ended his fifth day creation when he found the fish and birds to be pleasing to him. Some of the birds can now sing and eat mosquitoes, which so pleased the Lord.

Though species adaptations continue to this day, all the kinds of fish and birds were created by the end of the fifth day, which coincides with the disappearance of the dinosaurs, currently reckoned to be about 65 million years ago. The appearance of all the birds by this date is not currently appreciated due to the fact that the soft bones of small birds do not survive to be petrified.

Sixth Day of Creation

And God said, "Let the land produce living creatures according to their kinds: the livestock, the creatures that move along the ground, and the wild animals, each according to its kind." And it was so. God made the wild animals according to their kinds, the livestock according to their kinds, and all the creatures that move along the ground according to their kinds. And God saw that it was good. [Genesis 1:24-25]

The sixth day of creation marks what was long called the Tertiary period beginning about 65 million years ago and ended at the beginning of the

Quaternary which begins what is now supposed the Ice Age but should instead be recognized as the beginning of the antediluvian world.

Then God said, "Let us make mankind in our image, in our likeness, so that they may rule over the fish in the sea and the birds in the sky, over the livestock and all the wild animals and over all the creatures that move along the ground." [Genesis 1:26]

Thus, we see mankind's beginning only about 6,000 years ago!

57

The Earth, Climate, Animals, and Civilization Before and After the Flood

This is the account of the heavens and the earth when they were created, when the Lord God made the earth and the heavens. Now no shrub had yet appeared on the earth and no plant had yet sprung up, for the Lord God had not sent rain on the earth and there was no one to work the ground, but streams came up from the earth and watered the whole surface of the ground. [Genesis 2:4-6]

The above verses that immediately precede the account of the planting of the Garden of Eden, explain why the earth's earliest civilizations were based on the irrigation of river valleys. It is because at that time the earth had no rain!

Such rivers as existed before the Flood were fed by artesian springs from underground aquifers as was also the case of the Garden of Eden. These rivers emptied into the oceans causing the continents to steadily rise and expand due to the weight of the water being removed from the continents into the oceans. As the continents rose and expanded, shells formerly on the bottom of the oceans were elevated with the continents, some becoming part of the rising mountains. This took place over millions of years.

The Garden of Eden

Harvard University's James Sauer identified the Pishon River from satellite photos this river's dried bed of which flows from land of Havilah in Saudi Arabia and meets the Tigris and Euphrates Rivers in Kuwait.[60] The remains of the Garden of Eden as they may now be lie beneath the land of Kuwait or else under the Persian Gulf which formed when the waters of the Flood covered the valley where flowed the Gihon River from the Garden of Eden.

A river watering the garden flowed from Eden; from there it was separated into four headwaters. The name of the first is the Pishon; it winds through the entire land of Havilah, where there is gold. (The gold of that land is good; aromatic resin and onyx are also there.) The name of the second river is the Gihon; it winds through the entire land of Cush. The name of the third river is the Tigris; it runs along the east side of Ashur. And the fourth river is the Euphrates. [Genesis 2:10-14]

This land of Eden thus sank after the Flood due to the collapsing of the garden following the emptying of the reservoir which fed the springs forming the headwaters of the rivers, thus reversing the flow of these rivers whose ancient banks can be seen as having shifted due to being fed by new springs and rivers from the Zagros mountains. The Gihon River flowed through what became the Persian Gulf where, after the Flood, lived ancient Cushites prior to some of the Cushites migrating to southern India, Australia, and around the Arabian coast to the horn of Africa.

Nimrod was son of Cush, son of Ham. This mighty hunter was the "Daniel Boone" of the land of Shinar, which the Greeks would call Mesopotamia. As I mention in the *Archaeological Evidence,* a great change in the relationships between humans and animals followed the Flood. Ancient Near Eastern drawings show antediluvians feasting with wild animals who may have then lived in the same houses where the bones of some of these wild animals have been found. After the Flood, the Lord God put an end to such a close relationship between humans and wild animals.

Then God blessed Noah and his sons, saying to them, "Be fruitful and increase in number and fill the earth. The fear and dread of you will fall on all the beasts of the earth, and on all the birds in the sky, on every creature that moves along the ground, and on all the fish in the sea; they are given into your hands. [Gen 9:1-2]

This also meant that the wild animals suddenly became dangerous. Thus, we see houses just after the Flood being built with inner

---

[60] James Saur, "The River Runs Dry - Biblical Story Preserves Historical Memory," Biblical Archaeological Review 22, (July/August 1996), 52

courtyards, a way to protect humans from the animals as humans went about their daily activities. This also points out why the hunting skills of Nimrod were so important for establishing the first settlements in the first migrations after the Flood. These frontier settlements would grow to become the first cities after the Flood.

Before the Flood, what are now the earth's desert belts were lands of rivers and lakes and places of considerable human settlement. At the other extreme, what are now the arctic regions and colder northern lands as well as Antarctica were at that time places where lived elephants and other mammals that are currently found only in tropical or temperate areas. That means there was vegetation in these regions which we may suppose grew along water ways and on grasslands watered by dew. Cloudless skies as would exist during a time when the earth had no rain have a greater day/night temperature shift, which is conducive to great amounts of surface dew as is perfectly described in the verses quoted at the beginning of this chapter.

The temperature of the earth before the Flood can be determined by the size of the megafauna living before the Flood. Larger body mass is required to survive in colder climates, which minimizes the surface of an animal's skin or fur that is exposed to cold. Thus, the further north one travels, the larger are the animals. Contrary to what the Creationists claim, the large size of these megafauna indicates that the climate before the Flood was everywhere cooler than our current climate.

Even so, prior to the Flood, the arctic regions could not have then been covered with ice and snow. Otherwise, these animals would have nothing to eat. There was neither snow nor ice cover in the polar regions because at that time, there was no rain. Though there was sufficient dew for grasses, there was neither rain nor snow to create glaciers, whether on the tops of mountains or near the poles.

The rainbow which appeared for the first time after the Flood as a sign of God's promise that he would never again destroy the earth by water suggests that God also used the changes of the earth associated with the Flood to initiate the current hydrological cycle based on rainfall. Rainfall however required the creation of the earth's seasons. In Chapter 59, I will explain how this happened due to a sudden shift in the earth's tilt. Thus, the Lord promised,

"As long as the earth endures, seedtime and harvest,
cold and heat, summer and winter,
day and night will never cease." [Genesis 8:22]

This shift from irrigation to rain explains the mysterious prophecy from Noah's father that his son would give mankind rest.[61] Noah's name in the Hebrew language means *rest*. Rainfall eliminates the need for irrigation which leads to the need for slave labor. As alternative archaeologists such as Graham Hancock point out, the world's most ancient civilizations suddenly and mysteriously disappear. These ancient civilizations were built along river valleys. Likely, such great structures as the Pyramids of Egypt and even the towns of the American Southwest were built by slave civilizations which developed for the purpose of building and operating irrigation systems. Due to the land now being watered by rain, farming became less laborious since the Flood.

Today, the earth's rain originates from the evaporation of the earth's tropical oceans. Moisture-filled air is lighter than dry air. As moisture rises, it forms clouds that rise high in the skies and expand across the desert belts north and south of the equator. When these clouds reach the cooler temperatures of the earth's temperate regions, they drop their moisture as rain. The earth had no rain before the Flood because the earth's orbital tilt was either non-existent or else too small to give rise to hot summers and cold winters.

---

[61] Genesis 5:29

58

How Animals Crossed the Oceans Before and After the Flood

You are to bring into the ark two of all living creatures, male and female, to keep them alive with you two of every kind of bird, of every kind of animal and of every kind of creature that moves along the ground will come to you to be kept alive....Pairs of clean and unclean animals, of birds and of all creatures that move along the ground, male and female, came to Noah and entered the ark, as God had commanded Noah. [Genesis 7:19-20, 8: 8-9]

I have already mentioned the miraculous nature of the above verses. We do however have manifest evidence of this miracle, the change of fauna on every continent from before and after the Flood. I also mentioned the larger size of the fauna before the Flood which scientists refer to as Ice Age *megafauna*. This is not just the case in Europe, Asia, and the Americas, but also in Australia.

The animals which the Lord brought to the Ark would have been young and healthy specimens of these same megafauna. Their genetic offspring quickly adapted to the new climate and habitats after the Flood, forming the numerous species which now roam the earth. As occurs with selective breeding, geographical isolation develops new species at a rapid pace. The so-called genetic clock means little because when isolated species again mix, the genetic clock goes into reverse.

Species classification has long been a problem for naturalists. To solve this problem Johannes Buteo (1492-1572) developed the binomial Latin classification of species by trying to determine how many kinds of animals could fit into the Ark! Buteo used *species* for the classification of kinds and subdivided it by *genus*. During this era, the Latin *species* was sometimes regarded as being more general than *genus*. Buteo was followed by Athanasius Kircher (1602-1680), Bishop John Wilkins (1614-1672) and his assistant, the young John Ray (1627-1705) who tells us that 'after long and considerable investigation, no surer criterion for determining species has occurred to me than the distinguishing features that perpetuated themselves in propagation from seed.'

Evolutionists have an agenda to deliberately obscure the classification of species and thus define as species what are but geographical isolates. In doing so they obscure differences between species adaptations and what they claim are evolutionary production of new kinds of life. They do not even attempt to determine just how many species can produce fertile offspring with other species, the only way to understand how many animals that God needed to bring aboard the Ark. This has caused some to deny that adaptations happen sufficiently rapidly to explain the diversity of species following the Flood. As animal breeders know, it does not take many generations to create new breeds,

Most of what are now understood as species are either those which dwell in water and thus did not need to be saved by the Ark or else were kinds of microbes, insects, and invertebrate which would scarcely impact the space problem on the Ark. Thus, we need only consider the basic kinds of a few thousand species that are larger than a rabbit but smaller than a sheep and another several hundred kinds that are larger than a sheep up to the size of a giraffe or elephant. There is no reason to think that the one and a half million cubic feet Ark was particularly crowded.

Fig 54 – Line drawing of floating islands

Based on my analysis of archaeologically determined radiocarbon dates and artifacts, it seems that Noah built the Ark in what is now the Republic of Georgia where exists diverse ecological habitats between the Black and Caspian Seas.

As we can tell from prehistoric

archaeological ruins many of which lie just offshore from the continents, it seems that before the Flood the distribution of the continents was not so different from today, though many land bridges, as between Great Britian and the continent of Europe did then exist. Even so, the oceans as they existed after the Flood have impacted the distribution of animals. Many islands have few native species. Small animals however can raft on floating limbs while hurricanes can carry some of them across hundreds of miles of oceans.

Another means of voyage would have been the thousands of floating islands created by the flotsam and vegetation from forests uprooted by the deep and swift waters of the Flood. Depicted in Fig 54 is a line drawing of some of these islands which existed on the Atlantic coast of Africa until the early nineteenth century. Such floating islands not only preserved flora across the earth but also served as rafts for animals crossing oceans. Such rafts would have preferred tree dwellers and climbers as the case of most of the native animals of South America.

We have of course to consider the proverbial kangaroos of Australia by which many have mocked the account of Noah's Ark, how these managed to cross the Wallace Line which separates the fauna of Asia from those of the East Indies and Australia. These animals were surely guided and protected by their Creator as remains the case to this day. Marsupial animals however need not have been delayed by the weeks or months needed for young mammals to move and feed on their own. We may suppose that water barriers such as the Wallace Line appeared after the crossing of the marsupials. The dingo however would have been brought by the first Cushite families who settled Australia after the Flood.

59

Erratic Rocks from the Earth's Bedrock Propelled by the Swift Waters Bursting from the Great Deep

In the six hundredth year of Noah's life, on the seventeenth day of the second month—on that day all the springs of the great deep burst forth, and the floodgates of the heavens were opened. And rain fell on the earth forty days and forty nights. [Genesis 7:11-12]

The above verses also explain how the antediluvian earth then watered by springs, rivers, and dew changed to our current earth that is watered by seasonal rains. Contrary to the Creationists, as also Diluvium geologists such as DeLuc, Cuvier, and Buckland, the waters of the Flood did not come from the oceans but from the artesian springs of the Great Deep. Prior to the Flood, these springs fed the earth's rivers.

The Great Deep consists of water inside the earth which, as I explained in Chapter 56, exists in a plasma state due to the great pressures and temperatures at such depths. Some of this water rises though the earth's mantle to become the liquid water contained in great reservoirs beneath the earth's impermeable base rock. Most of the water in these reservoirs near the surface of the earth were emptied at the time of the Flood causing the earth to collapse into the resulting caverns. However, many of the emptied reservoirs turned into great underground caves.

The cooler waters released from these underground reservoirs into the warm oceans would have created great hurricanes, thereby opening the floodgates of heaven. There had never been rain on the earth when appeared a sudden downpour that lasted for 40 days. As the earth's temperature stabilized the rains ended. The water which continued to flow for five months came from the underground reservoirs.

THE SUDDEN TILT OF THE EARTH'S AXIS OF ROTATION

The earth's seasons are primarily due to the earth's tilt which causes the respective northern and southern hemispheres to face the sun during summer but tilt away from the sun during winter. This points to the tilt as occurring at the time of the Flood and as being the cause of the Flood. As I explained in Chapter 33, changes in the tilt are the favored explanation of the Ice Age, though assumed to have been gradual and regular.

I am not the first to suggest that God used a great comet to initiate the Flood. So did William Whiston, Isaac Newton's chosen successor to his chair of mathematics at Cambridge University. Though water from the comet would have been unnecessary to cause the Flood, the comet had to be of such size and closeness to the earth and on such a trajectory as to shift the axis of the earth's daily rotation from vertical to its current tilt. This sudden tilt of the earth would have broken and loosened the rocks forming the earth's bedrock, thus emptying the great underground reservoirs that had been supplying the earth's artesian springs.

EXPLAINING THE *ERRATICS*

The swiftly moving waters from the suddenly expanded springs of the Great Deep propelled the loosened bedrock to the places where these erratic rocks are found today. The sudden tilt caused great underground quakes that also led to depositing and mixing of freshwater shells from the earth's rivers and lakes with older marine shells as pointed out in 1808 by Cyprien Prosper Brard (c.f. Chapter 18).

The emptying of the underground reservoirs which fed rivers and lakes prior to the Flood caused a considerable rise in the elevation of land, currently being wrongly attributed to the melting of Ice Age glaciers. The emptying of the great weight of these underground waters also explains a sudden rising of the mountains at the time of the Flood, including new volcanoes. However earthquakes continue to occur as water continues to slowly empty from the continents whether from inland lakes or the virgin internal water expelled by volcanoes. Such rising of land explains the evidence of ancient vegetation which can only grow at low altitudes is also found at ancient archaeological sites on the plains just south of Mount Ararat which are now about one mile above sea level.

What makes the *erratics* important is that they are evidence and clues to the location of springs breaking forth from bedrock that once contained giant reservoirs of water as those under the plains of Kansas that continue to provide plentiful water for irrigation. However, most of the earth's current streams, which are fed by water tables replenished by rainfall, were created after the Flood. This explains the relatively shallow banks of most streams and rivers. Had these streams been flowing for hundreds of thousand or even millions of years, they would be far deeper, while eroded sand would have extended river deltas far into the oceans.

As to where to find the springs of the earth that emptied the waters contained in underground aquifers in such amounts as to cover the tops of the mountains, one need only follow the *erratics* to their geological origin. In the northern regions, this matches the supposed flow and source of the Ice Age glaciers. The geologists who developed what were successively understood as the *Diluvium,* then as *Drift,* and currently as deposits of *Ice Age* glaciers mostly studied the lands of northern Europe, northern America, and Canada which are currently identified as the lands covered by Ice Age glaciers. However, the same kind of unstratified deposits which cannot be explained by ongoing geological processes exist the world over. In more southerly and tropical areas these deposits are covered by increasing depths of topsoil and vegetation aside from places where erosion has revealed these rocks and deposits from Noah's Flood.

Forming, sorting, and depositing the Diluvium

Though I have given much attention to *erratics,* as did Charles Lyell and glacier scientists who supposed they had discovered alternative explanations for these displaced rocks, the Diluvium consists of far more components. These include much of the earth's soil and gravel lying just below virgin topsoil and also below the deposits or ordinary floods but above the earth's base rock. Likely, most of the earth's soil and rocks lying above bedrock was eroded by the swiftly moving waters from the springs of the Great Deep.

If you have witnessed swiftly moving flood waters or tsunamis, you can imagine just how diverse was the Diluvium. In addition to rocks and gravels, it contained soils of every type, not to mention vegetation, forests, and the corpses of both megafauna and humans including those remains and artifacts which led to belief in the great antiquity of man!

The Diluvium includes boulder clay, ordinary gravels, and vast deposits of loess that lie upon the world's most fertile plains, formerly recognized as having been deposited by water but which geologists have tried to explain as somehow deposited by Ice Age winds. Most geologists ignore deposits lying on or just beneath the topsoil. These include the red clay mixed with stones that cover much of my Piedmont North Carolina. This is because they have no way to explain them however valuable an understanding would be to land conservation and development.

As happens in all floods, the swiftly moving waters blended and sorted the Diluvium into various layers, from flotsam and loess at the top to heavy erratics near the bottom, much of the latter being deposited on bedrock. The heaviest soils, those with iron oxide though mixed with scattered rocks, were the first to be deposited. Most deposits of this red clay sank to the bottom of the oceans aside from that captured by valleys such as those of the Piedmont areas of Eastern America.

Most of the wood and vegetation which lay on or near the surface of the earth were burned by forest fires in the centuries following the Flood. Wood houses buried by the Flood were often incompletely burned along with their drowned occupants causing archaeologists to explain this phenomenon as "partial cremations"!

60

The Draining Waters of the Flood and Supposed Reversals of the Earth's Magnetic Field

But God remembered Noah and all the wild animals and the livestock that were with him in the ark, and he sent a wind over the earth, and the waters receded. Now the springs of the deep and the floodgates of the heavens had been closed, and the rain had stopped falling from the sky. The water receded steadily from the earth. At the end of the hundred and fifty days the water had gone down, and on the seventeenth day of the seventh month the ark came to rest on the mountains of Ararat. The waters continued to recede until the tenth month, and on the first day of the tenth month the tops of the mountains became visible. [Genesis 8:1-5]

As explained in the last chapter, the sudden tilt of the earth caused the Flood due to it breaking the earth's bedrock, releasing the reservoirs of waters which formerly reached miles below the surface of the earth. As these aquifers emptied, the interior of the earth collapsed, giving rise to various unconformities which cannot otherwise be explained.

Wind in the first verse of Chapter 8 might have been better translated as God's *Spirit* which healed and sealed the surface of the earth from the trauma of the Flood. Releasing the great weight of the waters expelled from the continents caused the continents and mountains to rise as the ocean basins sank and stretched due to the weight of this same water.

The great lakes of the world as those in the US, Lake Victoria in Africa, and other lakes in Asia and Europe were created by the Flood. The same is true of the Great Salt Lake and the Caspian Sea which existed before the Flood but expanded due to the waters of the Flood. This sinking of the ocean floors due to increased water continues to this day along trenches in the deep oceans. This is because water continues to drain from the continents, also drying up formerly inland seas such as the Caspian and the Great Salt Lake.

The weight of the waters of the Flood pulled down the continental shelves where are found the remains of antediluvian civilizations. The archaeological evidence of the migrations from Mount Ararat immediately following the Flood reveal that the waters of the Flood continued to cover lower lying lands for centuries after the Flood.

This stretching of the great Pacific Ocean explains the volcanic 'ring of fire' that surrounds the Pacific Ocean. Volcanoes arise as magma from the mantel makes its way to the earth's surface due to cracks in the bedrock of continental shelves. The cracks are caused by the stretching of the oceans which has been wrongly explained as due to movements of supposed rigid tectonic plates. Likely, active volcanoes date from the Flood with many such as Mt Etna in Sicily created immediately after the Flood.

The sinking and stretching of the oceans from the waters formerly inside the continents better explain the behavior of tsunamis than does the theory of plate tectonics. For example, undersea quakes caused by the sinking of the ocean basin explains why the waters along the coast suddenly withdraw before the tsunami appears. This is due to the sudden collapse of the ocean basin which registers as earthquakes associated with tsunamis.

One does not need such a theory laden explanation as plate tectonics to explain the separation of the continents. From the time they appeared, the continents of the northern hemisphere have gradually moved southward and separated due to the centrifugal force of the earth's rotation. The other factor affecting the distribution of the continents is due to the principle of isostasy, as the earth's continents move so that oceans and land gather to opposite sides of the earth. Thus, the Arctic Sea is opposed by the continent of Antarctica at the opposite pole. Everything currently explained by plate tectonics is better explained by the sinking of the oceans and the rising of the continents due to waters draining from the continents which began with Noah's Flood.

Imagined Magnetic Pole Reversals

This brings us to dating the earth on supposed magnetic pole reversals which, as I mentioned above, is currently being used to date supposed ancient Ice Ages. In 1906, the French geophysicist Bernard

Brunhes (1867-1910) discovered that the polarity of permanent magnetism in newly baked brick aligns with the direction of the earth's magnetic field. He discovered that the same thing happens with the cooling of lava from volcanoes. Measuring this effect on the sides of volcanoes caused Brunhes to suggest that the earth's magnetic field had occasionally reversed direction. Though largely ignored at the time, the same claim was made a few decades later by the Japanese volcanic scientist Motonori Matuyama (1884-1958). Such reversals in the earth's magnetic fields would be used to explain the variations in the magnetic fields due to the spreading of the floors of the oceans. Interestingly, the first change seems to be the same as the 650,000 year Pench and Bruckner Ice Age duration. Thus, these reversals are used to date supposed ancient Ice Ages.

These "reversals" of the earth's magnetic fields are but an assumption based on tiny variations in the measured magnetic field of the earth due to permanent ferromagnetism. Did it ever occur to anyone that these variations are found at just those place as the sides of volcanoes and near the sinking ocean floor trenches where one should expect changes in the positions of permanently magnetized rocks! Surely some have done so, but I have never found this matter even discussed in the current scientific literature.

Young Earth Creationists suppose that the separation of the continents took place after the Flood. Inspired by plate tectonics, they have developed what they call the *hydroplate* theory that explains the Flood and the breakup of the continents after the Flood. While the hydroplate theory is utterly speculative, so is the theory of plate tectonics. Both are no better than geological video games aiming to explain what happens in the world hidden beneath the surface of the earth.

According to the theory of plate tectonics, the surface of the earth is divided into rigid plates that are being slowly moved by conveyor belts of molten iron and nickel that circulate deep inside the earth. The earth's magnetic field is supposed created from these conveyor belts that are just below the ocean basins in the earth's upper mantel. For reasons not well explained, the direction of the conveyor belts is believed to occasionally reverse, resulting in the reversal of the earth's magnetic field.

THE TRUE SOURCE OF THE EARTH'S MAGNETIC FIELD

As a former professional electrical engineer, I know that moving charges do produce magnetic fields. But to produce an electrical current deep inside the earth, one must have a dielectric (i.e. electrical insulator) so as to produce an electrical potential. Have these scientists never learned that earth is everywhere at ground potential? Because the earth is conductive, it must everywhere remain at a single potential even when ground currents flow through the earth due to lightning strikes and defective power lines.

The earth does however have a dielectric, though not inside the earth. It is the atmosphere which lies below the ionosphere that consists of positively charged particles called *ions*. As the ionosphere is moved by the earth's turning, a magnetic field is necessarily created. To discover the direction of the earth's magnetic field, just point your fingers in the direction of the rising sun. Your thumb will point towards the earth's magnetic north.

The earth's magnetic field does not point exactly north. This is due to the paleomagnetism in the earth's crust that does slowly move, though not because of supposed plate tectonics! As I mentioned, the continents are slowly moved due to the rotation of the earth. The shifting continents explains what is called *magnetic declination*, which are slight changes in the direction of magnetic north at different locations. We also see small changes in the earth's magnetic field whenever the position of paleo-magnetized rocks are shifted as happens with earthquakes, rising volcanoes, and spreading sea floors! However, those changes in the sea floor magnetism cannot be caused by reversals in the earth's magnetic field which can only happen if the rotation of the earth shifted from east to west. I doubt that has ever happened.

This also refutes claims made by Young Earth Creationists that the earth's decaying magnetic field proves that the earth is very young. Scientists do not know the cause of a slight decay in the earth's magnetic field at certain places in the southern hemisphere. The cause is likely due to variable sun activity in the earth's ionosphere.

This is why we see great changes in the magnetosphere of the earth when the ionosphere is being disrupted by the plasma of solar winds. Interestingly, the sun does have magnetic pole reversals which correspond to the 8-14 year cycles of these solar winds which give rise to such things as disruptions of telecommunications and unusual northern lights. The average period of solar wind cycles is 11 years, the variations likely being due to the position of the earth at the time these magnetic reversals of the sun take place.

As I just explained, magnetic fields are generated by the moving ionosphere of the earth. This is also true of the planets though varying with the different intensities of their ionospheres and their various rates of rotation. However, the movement of the sun's ionosphere is due to the gravitational attraction of the planets, which is almost entirely that of the massive planet Jupiter. Why have astrophysicists failed to notice that the solar cycle matches Jupiter's 12-year circulation around the sun, less an average of one year difference in measurements caused by the earth's position around the sun at the time of measurement?

From the era if the seventeenth century Scientific Revolution, many have speculated on the cause of the earth's magnetic field. Discovering the cause should have accompanied the discovery of the ionosphere early in the twentieth century before the impossible current explanation of the earth's magnetic field was proposed. While an impossible cause of the earth's magnetic field has long been accepted, the obvious cause seems to have never been considered.

PART VII

A NEW SCHOOL OF ORIGINS

61

SUMMING UP

The preceding chapter will end this review of the lack of serious thinking and needful skepticism in the current sciences which otherwise would have revealed serious problems including violations of the most basic laws of physics. I have in mind the impossible powers imagined for supposed Ice Age glaciers and overlooking what should have been an obvious solution to something so basic as the source of the earth and the sun's magnetic fields. To have supposed early man as being so excessively religious as to devote his entire production to constructing elaborate tombs shows just how deceived were those responsible for the current scientific narrative.

The mathematician Georg Cantor explained how such teachings have been so long sustained:

A false conclusion once arrived at and widely accepted is not easily dislodged and the less it is understood the more tenaciously it is held.

It should be clear that most who practice and teach the historical sciences do not well understand what they have been taught to practice and teach. One aim of this book is to help you understand these teachings better than most who teach them, at least until the latter read this book. For anyone seeking truth, this will be sufficient to reject much of what our modern world unconsciously supposes as truth. All that is required to reject the current scientific narrative is to understand it.

Even as a Christian, I did not doubt evolution until I learned its details. It is not the failure to understand the theory of evolution which causes one to be skeptical. Rather, it is the lack of knowledge about its foundational claims that sustains its hold over those who suppose themselves enlightened. Believing themselves well educated, many become arrogant, regarding themselves as a cultural elite. However, being humble about what we suppose we know is essential for discovering truth.

The American philosopher C.S. Peirce reminds us that scientific views are hardly challenged until someone has a reason to do so. From my youth, I loved the sciences because they do help us understand many things. However, as I soon discovered, not everything! Though sincerely seeking answers, my questions upset some of my teachers and when that happened, my classmates as well. No one likes being unpopular. This had nothing to do with how such teachings may have challenged the Bible. My aim was to understand what I was being taught rather than pretend to understand as would most of my classmates.

It was trying to understand what others did not recognize as problems that led to my contributions in the field of telecommunications, which included bringing the digital network to local exchanges at a time when the scientists of America's monopolistic carrier taught as being premature. It seems that financial motives were impacting their scientific judgments.

Not until I obtained faith in Jesus Christ did I have a reason to seek answers for what appeared conflicts between the Bible and the historical sciences. The answers turned out differently from what I expected but also from anything ever taught me.

Such problems however are not restricted to the secular sciences. In their different ways, both Old and Young Earth Creationists have built upon the secular narrative, trying to bring it in line with their interpretations of the Bible. Creationists make their own errors that blind them to truth as the case of Old Earth Creationists claiming that the Bible teaches a local Flood when it clearly does not and Young Earth Creationists claiming that while the Flood left massive evidence of dead fish, it somehow obliterated the artifacts, houses, and hard bones of antediluvian humans.

Both schools of Creationists believe in the imaginary Ice Age which shows just how much they are beholden to the current scientific narrative. Creationists can hardly dislodge a narrative of origins which in different ways is built into the foundations of their supposed biblical understanding. These supposed solutions have only hindered the search for better answers by those who love truth. Rather than dogmatically assert these supposed solutions, had we humbly admitted that we lacked answers, the problems may have been solved long ago. While bad answers block the search for better answers, humbly acknowledging questions for which we have no answers is no barrier to faith. Rather, that is how we obtain truth. Whether unbeliever or believer, know that

What keeps us in darkness is not because of what we do not know or understand. Rather, it is due to believing things that we have never thought to question but which happen to be false.

This kind of darkness blocks and dismisses light. Light cannot penetrate such thick darkness, why humility is so essential for obtaining truth. When we discover our mistakes, we should be bold to proclaim them. I trust that I have shown how seriously using the light of the Scriptures uncovers many scientific falsehoods. Know however that

The Scriptures are just so many words until we turn on their light by believing them sufficiently to use them in searching for truth.

62

SCRIPTURE SCIENCE,

A NEW SCHOOL OF ORIGINS

It is one thing to recognize problems in the current narrative of origins, but a far more difficult thing to provide a good alternative. It is however necessary that I differentiate the narrative by which I understand the origins of the world and mankind in the light of the Scriptures from those of the Creationists. I do so by referring to the understanding presented in this book as *Scripture Science* to distinguish it from Old and Young Earth *Creation Science*. That *Scripture Science* is an appropriate name should be clear from the multitude of Scriptures that have served as my light for understanding the sciences.

Table III on page 243 compares how origins are taught by four current schools embraced by evangelical Christians, also comparing them with Scripture Science, which becomes a fifth school of origins.

1. The school described in the first column is *theistic evolution* which is scarcely different from mainstream science except perhaps asserting that evolution was divinely guided. Theistic evolution, also called *evolutionary creationism,* is embraced by the organization *BioLogos* founded by Francis Collins who is famous for his role in sequencing the human genome.

2. The school described in the second column is known either as *Gap Creation* or as *Ruin and Restoration*. Taught in the notes of the influential Scofield Reference Bible, this was the most popular school of origins embraced by Fundamentalists in the first half of the twentieth century and more recently by prominent evangelicals such as the late Pat Robertson and the Southern Baptist leader Dr. Richard Land, who shared his view of origins with Norm Geisler and

myself. This view continues to be held by many older Dispensationalist theologians.

3. The third school listed in this table are day-age Old Earth Creationists, prominently represented by my friend, the astrophysicist Hugh Ross and the organization *Reasons to Believe (RTB)* that he founded.

4. The fourth school, Young Earth Creation Science (YEC), has the allegiance of most evangelicals today and is taught by a number of organizations, the largest being the *Institution for Creation Research* (ICR) founded by the late Henry Morris, the Australian organization *Creation Ministries International (CMI),* but today most prominently in America by my friend Ken Ham's *Answers in Genesis(AIG)* and his *Ark Encounter* in Kentucky.

5. The last school listed is my new school of Scripture Science (SS) which is the only one of these schools ever endorsed by the late pioneer of evangelical apologetics, Dr. Norman Geisler.

There is a school of interpreting Genesis known as the *Framework Hypothesis*. I do not regard this as a school of origins because the Framework Hypothesis does not suppose the Scriptures as teaching anything about the physical world. Rather, it is only for interpreting the Scriptures. There is also what might be referred to as the school of *Intelligent Design (ID)* represented by the *Discovery Institute* in Seattle, Washington. I do not compare ID as a separate school because it is embraced by every school of origins aside from those who teach evolution by unguided natural selection.

Perhaps, the most surprising difference between Scripture Science and the other schools concerns the age of the earth which I understand to be older than the earth's rocks, currently estimated to be about 5.4 billion years. As I understand the Creation account, these rocks were formed from trace minerals in the waters of the earth's creation. The entire universe is currently supposed as being about 13.7 billion years, give or take about 200,000 years. Know however that I have no strong views concerning the age of the universe, the age of the earth, or whether most rocks were created by one or more floods. If not convinced by their solutions, I value the skepticism of the Young Earth Creationists concerning current geological teachings.

Assuming that for most of its early history the earth existed in an aqueous state, I hold to the conventional dates of the earth's rocks until I find some reason not to do so. Due to paleontologists having found dinosaur remains with soft tissue, I question whether they can be so old as presently believed. Young Earth Creationists have questioned this and so should mainstream scientists! Scripture Science shares with the Young Earth Creationists an understanding of rapid speciation which evolutionists deny but which does occur when species become geographically isolated.

Scripture Science differs from all other schools of origins by denying there was ever an Ice Age and by holding to the water-based geology and cosmology taught in the Scriptures. Most important, Scripture Science excels all the other schools including mainstream science in its great simplicity, testability, and useful predictions. Because of this, Scripture Science will prevail over mainstream science.

Table IV on page 244 further compares Scripture Science with the only other school or origins teaching a worldwide Flood, that of the Young Earth Creationists. Note that while Young Earth Creations claim the pre-Flood world to have been warmer than today, Scripture Science teaches the antediluvian world to have been cooler and drier. While evidence of the victims of the Flood provided by Young Earth Creationists are mostly fish, Scripture Science identifies the massive evidence of antediluvian civilization and land animals that perished in the Flood. Thus, Scripture Science far better agrees with the Bible especially with regard to Noah's Flood.

Unlike all the other schools of origins, Scripture Science is able to identify the humans and animals who perished in the Flood as well as trace the historical nations to the remains of Noah's Ark on Mount Ararat and thus demonstrate that these remains are indeed those of Noah's Ark.

TABLE III - FIVE SCHOOLS OF ORIGINS EMBRACED BY EVANGELICALS

Scientific School	Theistic Evolution (TE)	Ruin & Restoration (Gap Creeation)	Old Earth Creationism (OEC)	Young Earth Creation Science (YEC)	Scripture Science (SS)
Supporting Evangelical Institutions	BioLogos	Some Baptist and Pentecostal Seminaries	Reasons to Believe	Institution for Creation Science, Answers in Genesis, CMI, many others	None
Prominent Defenders	Francis Collins, N.T. Wright, John Walton	Scofield Reference Bible, Pat Robertson,Richard Land	Hugh Ross, Walter Kaiser	Henry Morris, Ken Hovind, Ken Ham	Philip Williams, Norm Geisler
Extent of Flood	Local if even historical	Eiither Local or Worldwide	Local	Worldwide	Worldwide
Noah's Ark on Mt. Ararat?	Impossible	Possible	Impossible	Very unlikely	Very likely
Evolution (macro)	Accepts	Rejects	Rejects	Rejects	Rejects
Date of Adam	No historical Adam	c. 6,000 bp	c. 300,000 bp	c. 6,000 bp	c. 6,000 bp
Date of Flood	Borrowed ANE myth	c. 2349 BC	c. 100,000 bp	c. 2349 BC	c. 2349 BC
Age of earth	4.5 billion years	4.5 billion years	4.5 billion years	c. 6,000 years	Older than 4.5 billion yrs
Age of universe	13.7 billion years	13.7 billion years	13.7 billion earths	c. 6,000 years	13.7 billion years
Sun, moon, planets, stars	Nebular hypothesis	Nebular hypothesis	Nebular hypothesis	Created 4th 24hr day	Created on 4th divine day
Fossil-indexed geology	Deep Time	Deep Time	Deep Time	Created by Flood	Deep Time
Cosmology	From Big Bang	Genesis 1:1-2	From Big Bang	From 6-day Creation	From separation of waters
Archaeology	Follows mainstream anthroplogy	Follows mainstream archaeology	Follows mainstream archaeology	All human remains date from after 2345 BC	Evidence of violent antediluvians, postFlood spread from ANE
Era of Ice Age	2.5MY–12,000bp	After Flood	2.5MY–12,000bp	2345BC – 1500BC	Was never an Ice Age
Useful predictions	Few	Almost none	Few	Almost none	Many in every field
Earth's core	Iron/nickel	iron/nickel	Iron/nickel	Iron/nickel	Water in plasma state
Interior of stars	Hydrogen/Helium	Hydrogen/Helium	Hydrogen/Helium	Hydrogen/Helium	Water in plasma state
Theory of continents	Plate tectonics	Break up after the Flood	Plate tectonics	Hydroplate/other theories	Oceans sink/spread due to Flood
Ockam's Razor Test	Fragmented, muddled	Extremely muddled	Fragmented, muddled	Super magical	Integrated, elegant, powerful
Biblical Hermeneutics	Science is truth, Scripture is allegory	Scofield Reference Bible interprets Bible in harmony with science	Interprets Scripture to harmonize with science	Young Earth Creation Science interprets the Bible and Science	Jesus interprets Scripture, Scripture interprets scientific evidence
Testable?	Not very	No	Not very	Not at all	Very

TABLE IV - TWO SCHOOLS OF ORIGINS TEACHING A WORLDWIDE FLOOD

	Young Earth Creation Science	Scripture Science
Founder of scientific school	George McGready Price (early 20th cent')	Philip Williams. (early 21st century)
Well-known advocates	Henry Morris (ICR), Ken Ham (AIG)	Norm Geisler
Biblical hermeneutics	Historical literalist	Historical literalist
Date of Adam, first man	c. 6000 years ago	c. 6000 years ago
Date of Flood	c. 2345 BC	c. 2345 BC
Creation days	Same duration as man's 24-hour days	Creator's days (evening and morning)
Age of earth	c. 6000 years	Older than 4.5b year old rocks
Age of cosmos	c. 6000 years	Possibly 13.4 billion years
Extent of Flood	Worldwide	Worldwide
Animal death before Adam's Fall	No	Yes, God never offered animals eternal life
Geological evidence of Flood	Fossils, all fossil fuels, sedimentary rocks	Diluvium renamed Pleistocene ("Ice Age")
Was there an Ice Age?	Yes, after Noah's Flood	Neither before nor after Noah's Flood
Climate prior to Flood	Much warmer, tropical	Cooler, streams from earth but no rain
Evidence of humans who perished in Flood	None, Flood vaporized all remains of humans/cities/artifacts	Numerous remains, artifacts of drowned antediluvians worldwide
Evidence of violence before the Flood	None	Massive evidence of violence worldwide
Evidence of victims of Flood	Fish (>90%), dinosaurs,extinct fossils	Humans, cities, megafauna, birds
Resting Place of Noah's Ark	In mountains of ancient Urartu	Top of highest mountain in ancient Urartu
Do remains of Noah's Ark still exist?	Probably not	Probably discovered
Did Flood destroy Tigris/Euphrates Rivers?	Yes, completely	No, though modified by Flood
Archaeological spread from Ark?	No evidence	Yes, traced to Mount Ararat
Evidence of Ancient Babel	Not identified	Uruk, city established by Nimrod in Shinar
Evidence of dispersion of Noah's family	From genetics	New dispersion mapped from Near East
Archaeological evidence of Flood	No archaeological explanation!	Corrects mainstream archaeology
Geological evidence	Reinterprets Deep-Time geology as Flood	New water-based geology
Astronomical and astrophysical evidence	Compresses mainstream timescale	Radically-new water-based cosmology
Evidence of 1st-Day light	No	Yes, cosmic background radiation
Evidence of 2nd-Day separation of waters	No	Expanding waters create cosmic space
Evidence of 3rd-Day plants	No, plants created fully grown	Oil and natural gas
Created on 4th Day	Sun, moon, and stars	Sun, moon, and stars
Created on 5th Day	Fish, birds	Fish, dinosaurs (early birds), current birds
Created on 6th Day	Dinosaurs, land animals, Adam and Eve	Land animals, Adam and Eve
Makes testable/useful predictions	Few if any	Numerous in every historical science
Simplifies scientific understanding?	Creates many new mysteries, solves few existing mysteries	Simple but elegant and powerful explanations

63

Peer Review and its Limitations

Some wish to know what has been published about this discovery in peer-reviewed scientific papers. There have been a few papers including some of my own that have been presented at scientific conferences held in the Republic of Türkiye, the Republic of Georgia, as well as in Hong Kong, which is now part of the Peoples Republic of China. These papers of various scientific merit were published in the proceedings of these conferences. But scientific publications from an archaeological survey and excavation of the sites on the mountain must wait for the Government of Türkiye to issue an archaeological license.

The current academic system is built on papers accepted by scientific and academic journals of various levels of prestige, especially how frequently they are cited in other papers. This process goes back to the international community of letters written in the era of the Scientific Revolution and the journals to which these letters were written and published, some of which I mentioned in Part III. As the fields of knowledge expanded, these journals became ever more specialized to the point that ever fewer experts are now interested if they can even understand papers written to various specialized journals. Experts are themselves but laymen in scientific fields other than their own even if some of the other fields may be necessary for understanding their own field.

As those who write these papers know, these journals have become a business, as indeed have the universities and institutions themselves. The same is true of the grants that fund the sciences. Universities obtain a cut of these grants and thus make decisions about their faculty that are frequently based on the potential for revenue. In many cases, advancement of an academic career will trump judgment based on truth and merit. Those who are interested in important questions pertaining to science and history lament this situation, but what in our present world has not been reduced to a business?

A problem is that this matter of Noah's Ark and Flood requires expertise in a multitude of specialties of which no one person possesses. Who are the peers with expertise in reviewing claims for Noah's Ark? Aside from those who participate in the conferences mentioned above, where are the Noah's Ark experts and how do we recognized them among those specializing in investigations of Noah's Ark? If archaeologists and scientists have been blind to the manifest evidence of Noah's Flood, how can they judge the evidence of Noah's Ark?

This is not to say that specialize papers are not to be desired, especially by this author. Thus, I hope to launch a new journal pertaining to Noah's Ark and Flood. Remember that there are no uninterpreted facts. Proper scientific judgments must be proceeded by a scientific understanding of the waters which floated the Ark to the top of Mount Ararat, the purpose of this book.

64

A Paradigm Shift

The matters discussed in this book have been presented as a new paradigm, thus requiring a shift either from the current paradigm of mainstream science or perhaps from that of Young Earth Creation Science. A paradigm shift is a new way of interpreting old scientific facts. It is putting the current scientific "facts" in a new light. Judging paradigms is different from judgments in the peer-reviewed papers that take place in the context of a prevailing paradigm. Thus, mainstream scientific journals do not ordinarily accept papers from those committed to Creation Science? Journals devoted to Creation Science do not often publish papers presuming the truth of evolution.

Those beholden to the current paradigms, whether those of mainstream or Creation Science, may not even understand what is presented here. As the philosopher of science Thomas Kuhn famously explained, judging paradigm shifts is not normal science. It is what Kuhn called *revolutionary science*.

Judging paradigm shifts concerns the matters in the last 3 rows of Table III and the last 2 rows of Table IV. The criterion called *Oakham's Razor* amounts to maximum simplicity which comes through unifying things that are not well integrated in the light of the current paradigm.

For example, Scripture Science unifies archaeology and geology by showing how most prehistoric burials were created by Noah's Flood, the evidence of which the current paradigm has supposed as being instead evidence of a long Ice Age. The paradigm of Scripture Science also unifies the laws governing cosmological bodies, from comets and meteorites to terrestrial moons and planets, up to giant planets and stars. Scripture Science also has a simpler geology in which water, granite, and Noah's Flood are more fundamental and useful for understanding the earth than are volcanoes and plate tectonics which serve as the basis of the current paradigm.

Of greatest importance is a paradigm's ability to predict things that no one previously considered. In the case of Scripture Science, this includes (1) the depth of burial of prehistoric sites based on terrain and altitude, (2) the southern latitudes of erratic rocks, (3) the behavior of tsunamis, (4) the source of the magnetic fields of the earth, planets, and the sun, and (5) the migrations of the world's nations from Mount Ararat!

The third way of judging a paradigm shift is specific to Christians. This is biblical hermeneutics, whether relevant texts in the Bible are being used as the light for the scientific data (i.e. a literal versus allegorical interpretation), whether the understanding of the relevant texts are interpreted so as to conform to the science of a particular paradigm (e.g. so as to teach a local Flood), or whether the paradigm imposes some external light on the biblical text as in the case of the Young Earth Creationists imposing a 24-hour duration on the days of Creation. While Old and Young Earth Creationists engage in different kinds of eisegesis, Scripture Science uses the light of a contextual reading of the Scriptures. I trust that you will note that Scripture Science is remarkable for its use of so many biblical texts that no one has previously used for explaining scientific matters.

65

JUDGING OLD AND NEW NARRATIVES

A better term than *paradigm* for describing a new school of origins is the word *narrative* which has recently become popular for orienting facts to particular political or ideological commitments. However, the only way to describe historical events is through narrative. Scripture Science is not so much a new narrative as the extensive and thorough use of the narrative of beginnings put forth in the Bible. This new narrative is thus based on the oldest of narratives. Because a narrative of origins impacts history, I refer to *Scripture History* as well as *Scripture Science*. Scripture History is understanding world history in the light of the Bible.

Confident as I am concerning this new narrative of origins that is based on seriously, thoroughly, and consistently using the light of the Scriptures, I am not infallible. However difficult and in whatever forum, everything presented here should be severely and robustly tested, from the most consequential matters to historical mistakes that may have little or no impact on the larger matters addressed here.

I have this maxim for judging truth.

Truth should be judged by how proponents address what may seem the most difficult questions in public forums open to opposing views and which respect the dignity of all.

Note that this is not something that we often see. While those with something to hide or those who are not prepared to defend their views will avoid serious examination, the authentic does not mind being tested. Thus, I look forward to addressing what may seem the hardest questions. I only ask those who wish to do so to carefully read this book. So long as questions are asked in a manner respectful of all, I look forward to addressing whatever might be unclear or perhaps unsound. Correcting such errors as might be present will lead to an even better understanding, which is how I arrived at what is presented here. I arrived at these teachings by examining relevant scientific evidence in the light of the Scriptures, by consistently applying the fundamental principles of physics, and by investigating how the views that I reject became part of the scientific narrative.

The primary examination of this new narrative can hardly take place in specialized academic journals but open and before the world as when the current scientific narrative was established, that is by books such as Lyell's *Principles of Geology* and Darwin's *Origin of Species,* in public lectures, popular journals, and magazines more than in specialized scientific journals that few are able to read or understand. Unfortunately, the current narrative involved a shift to relying on specialized experts, which is also how the current scientific narrative has been preserved. The Western world shifted from trusting the Bible and theologians to trusting a secular priesthood referred to as *scientists.*

Such important questions should never be left to supposed experts! In any case we must ask, just who are those who specialize in origins? Who is experienced at criticizing explanations of origins as the case of the 5 schools recognized by various evangelicals? Who might do so without an agenda other than the search for truth? Is that forum some theological or scientific journal? Should that be a secular or religious journal?

Those engaged in a specialized field are recognized as peers by those working and publishing in the specialized field. Thus, we expect peer review of any writing that challenges any aspect of the current science. However, peer review is also the priestly duty of Christian leaders and scholars. It dates at least to the testing of prophecies, as advocated by the Apostles. We see peer review in Peter Abelard's 'Sic et Non' (yes and no) by the founder of one of the world's first universities and in the review of writings among the unofficial republic of letters which arose among Christian scientists during the Seventeenth Century, most of them ordained ministers. As I mentioned in Chapter 63, this eventually led to edited journals.

The earliest thinkers who shared and reviewed letters and books were of wide erudition, as the case of the major thinkers of the Renaissance and the Scientific Revolution. Due to the problems discussed in this book, wide erudition has almost disappeared! Regardless the pretentions of progressives, erudition is becoming ever more rare. This is because strong Christian faith is necessary for preserving truth.

Just what journal today reviews challenges to the larger scientific narrative that cannot be understood within the narrow bounds of some specialty? The problem of specialized knowledge is that with regard to true expertise, there is no such thing! Even in special field of telecommunications, I discovered that there existed an abundance of more specialized fields. I also discovered that no one can truly be expert in any of the various fields of telecommunications unless he possesses the fundamental knowledge of all related fields. How much more the case for the expertise needed for understanding the origins of the cosmos, the earth, or man himself! It is the reason that the many problems I point out haven't been investigated if even noticed by ordinary workers in the numerous sciences. By continuing to work under the current scientific narrative, this situation will only get worse as specialists study ever more specialized questions which can be understood by an ever smaller number of people.

Archaeologists are sometimes supposed as an exception to specialization. Indeed, they should understand the many fields and ways of testing used in their profession. They also need to be historians, at least for the times and places pertaining to the sites they investigate. But few archaeologists have expertise in recent earth geology as also in the anthropology from which the current prehistory derives. Even fewer glacier scientists have expertise in archaeology, though the Ice Age dating began with archaeological dating! With regard to the origins of man, the blind led the blind as the historical sciences fell into the Ice Age ditch.

If most prehistoric graves were created by the Flood of Noah, who has the expertise to judge? How can experts who failed to notice the manifest evidence of the worldwide Flood of Noah or who failed to notice the new migrations that began on Mount Ararat judge the evidence of Noah's Ark? That of course should be the job of the most accomplished archaeologists, those who understand the historical development of archaeology as I explained in Part IV. But this will require that they do so in the light of the new paradigm.

Archaeologists are already interested in the ancient wooden structure on Mount Ararat, if only for the fact that these remains get the world's attention. Some of the world's most expert in archaeological dating are eager to test whether, as I claim, the artifacts found in this ancient wooden structure match those of the migrations which spread from this mountain. Because I have investigated the ancient world from the evidence and theories of mainstream archaeology, mainstream archaeologists and scientists better understand what is written here than either Old Earth or Young Earth Creationists, both schools being seriously deficient in the study of prehistoric archaeology!

Discovering and testing for truth require that it be presented so as to be tested by all who love and seek truth, who recognize our limitations and who understand that there remains much truth that we do not currently possess. The more one learns, the more one recognizes how little we know, but also how much we can learn should we be hungry for truth and humble enough to receive it from where we might not expect?

66

New Centers for the Study of Origins

I can only summarize here the needs for two new scientific centers. Other centers might also be established at whatever existing or new institutions that might wish to focus on specific areas and fields as mentioned below.

A Scientific Center at Mount Ararat

If the ancient wooden structure on Mount Ararat is the place of origins of all the world's animals and humans, there is needed by this famous mountain a scientific center for the study of origins: the spread of nations, human arts, plant and animal domesticates, birds and land animals. This region is known as having the world's earliest viniculture, domesticated horses, sheep, goats, and grains and other domesticates as should be expected due to departures from the Ark.

Here is an opportunity to obtain a true understanding of species, which is useful for wildlife preservation and land development. DNA studies of samples from the sites on this mountain and of wildlife near this mountain become the basis of such studies.

In partnership with certain Turkish universities and the Ministry of Culture and Tourism of the Republic of Türkiye, including the Directorate of Culture Assets and Museums and the Ministry of Mines, the Discoverer Ahmet Ertugrul has established such a center near the south side of Mount Ararat. Called *Noah's Village,* the center is in direct sight of where the broken parts of the ancient wooden structure lie buried high on the mountain. Already, this center is hosting conferences sponsored by Turkish universities and attended by archaeologists, scientists, and culture historians the world over. This center however needs a laboratory suitable testing samples and artifacts from the ancient wooden structure on top of the mountain. All this requires funding, most of which may come from America.

A New Center for the Science of Origins and World History

No less important than the scientific center by Mount Ararat is an online center for research and scientific investigation and for teaching a new generation the origins of the world and the ancient history of mankind in the light of the Bible. American Christians, help me fund and build this new center for the Science of Origins.

This center will host:

1. A new school of *Scripture History* (world history in the light of the Scriptures) teaching American natives how and when their ancestors who descend from Noah's son Shem migrated across Asia and the Pacific Ocean to the wilderness of what Europeans would discover as a New World. The school will likewise teach the origins of natives from every continent. This will be part of a new course in the history of nations beginning with those of antiquity through how new nations became established in the post-colonial world. Students will learn how to identify the spearpoints and fortifications from the advanced but violent civilizations who perished in the Flood even in my home state of North Carolina. Properly understanding the history of native Americans means distinguishing their ancestral temple burials from the violent people who lived in America in what archaeologists currently understand as *Paleolithic* and *Archaic* America but in fact antediluvian America.
2. A new school of American History which will teach how the exploitative colonial culture of America was transformed by those seeking religious freedom in the wake of the back-to-the-Bible Reformation, how the Bible served as the central text from the colleges to the grammar schools, thus how colonial America became the world's most literate and equitable society, becoming the world's first self-governing nation.
3. A new school of Religious History that teach the Platonist roots of Eastern religions imported to India by Alexander the Great. They will learn that the Indians did not teach Plato, as many supposed, but the Greek invaders were overlords who taught the Buddhists and Hindus, the latter being the most recent of India's religions formed from India's polytheism. It will also teach how Christianity has influenced contemporary Hinduism.

4. A new school of *Scripture Economics* to show how the Industrial Revolution depended on middle class consumers as existed in Early America which aimed for its people to forgo buying imports in order make their nation self-sufficient and strong. Far from being individualists as wrongly taught today, early Americans depended on their churches and communities. While the classical pagans looked down on manual work, when it comes to serving others, Christians have not. People's needs rather than their greed should inform a nation's economic policies. However, as the Scriptures advise us, those who refuse to work should not eat at the public expense. Such a rule allowed Jamestown to become the first English settlement in America.

5. A new school of *Scripture Anthropology* which will give the deathblow to racism by exposing its philological and scientific roots as I explained in Chapter 35. This will replace the scientific narrative of man's origins established on racist teachings by showing how all nations have their origins as one family who lived in the ancient wooden structure that I visited on Mount Ararat. Scripture Anthropology will teach the spread of arts and agriculture from Mount Ararat.

6. A new school of *Scripture Geology* based on the waters of the Flood and the waters inside the earth as outlined in Part VI. The geological survey of most of the world's countries was conducted in the light of an imaginary Ice Age. This meant neglecting the most recent geological deposits that are important for land development and wildlife. Understanding the origins of the soil and rocks that lie on or just below the ground will greatly assist this task as also a better understanding of streams and rivers.

   Geology, which was the first science in which ordinary people could participate, should return to its democratic beginnings. There is no more enjoyable outdoor activity than doing geology while exploring the great outdoors. This center will organize these surveys and teach a new generation of geologists how to report the results. Scripture Geology replaces basalt with granite for understanding the earth's fundamental bedrock and will also replace plate tectonics with an understanding of changes to the earth brought about by the rising and sinking waters of Noah's Flood. The center will publish a revolutionary geological text.

7. A new school of *Scripture Cosmology and Astrophysics* to work with astronomers and astrophysicists for unifying and advancing our

understanding of cosmic bodies, from asteroids to super galactic clusters through a new water-based cosmology as outline in Chapter 56.

8. A new school of *Scripture Archaeology* as outlined in Part II to work with archaeologists and archaeological organizations concerning world prehistory and understanding prehistoric sites in the light of the Flood.
9. A new school of *Scripture Biology* as outlined in Chapter 58 to partner with the new Scientific Center in Turkey pertaining to the ancient wooden structure and worldwide migrations of domesticates and wild animals from that region.
10. A new school of *Scripture Philology* and historical linguistics, including the study of languages, scripts, and literacy in the light of the Scriptures along with a Scripture-based historiography.

A New Encyclopedia of Scripture Science and History

As the original encyclopedias developed during the Enlightenment of the eighteenth century, an entirely new understanding of the world and its history requires a new encyclopedia. The advantage of an encyclopedia is to be able to independently develop and update the various topics covered in this book. We refer of course to an online encyclopedia made convenient by tools such as Wikipedia. This present book serves as the outline of the new encyclopedia and becomes the core of a new standard curriculum based on Scripture Science and History.

67

A New Standard Curriculum based on Scripture Science and History

Abraham Lincoln has been credit with saying that the philosophy taught in the classroom in one generation will be the philosophy of the government in the next generation. In Lincoln's day, *philosophy* encompassed what we now call *science* with its narrative of origins including ideologies based evolutionary science. Thus, we understand his statement. We have government policies that are the results of revisionist history taught in the universities decades ago but now being taught in primary schools. This revisionist history was due to the unanswered scientific challenge to the Bible and has undermined Americans' knowledge of the Christian foundations of America. In Chapter 80, I will discuss how this discovery can change the world by teaching the truth concerning the past. This will encompass a new standard curriculum based on Scripture Science and History.

Fortunately, classrooms are becoming online with the ability to widely and rapidly change whatever is taught and believed. It is no accident that changes in technology have assisted earlier world-changing messages as the case of (1) writing invented just in time to preserve knowledge of the world before the Flood, (2) the alphabet invented just in time for ordinary Israelites to record parts of the Law of Moses on their doors and cloaks, (3) the codex (i.e. the book) which appeared just in time for the gospel, (4) the printing press invented just in time for the Reformation, and (5) newspapers pioneered by publishing news of the Great Awakening revival. Whatever its problems which are serious and many, the Internet and the worldwide web are a Godsend.

In America's beginnings, the Bible was the chief text for education from the colleges to the grammar schools. The schools and most universities were established and operated by churches. Local and state governments eventually funded the schools though initially teaching a curriculum still grounded in the Bible. However, with the rise of evolutionary science, the schools became increasingly secular and based on a materialist understanding of society including moral relativism. Due to not being based on the Bible, the curriculum of specialized fields lacked connection to the history of ideas as was formerly the case of a liberal education of which every leader and teacher was presumed to know.

However, I strongly reject the trend of some Christians to idealize a classical education. A classical education is of course valuable and superior to the postmodern nonsense currently being taught. However, they do not seem to know that the men of the Enlightenment used their classical education to challenge Christianity and the Bible? If 'Western' refers to the Christian West, this curriculum was also based on the pagan civilizations of Greece and Rome. This includes the literature of the Romantic era that not only introduced racism but also the science and history that rejected the Bible. The classicalism of the Enlightenment also served as a defense for colonialism and slavery. Proponents of classical education should know that a totalitarian state is rooted in the teachings of Plato! The Greek aristocrats wanted freedom to do as they pleased but they did not allow teachings that challenged their own. They would have opposed America's First Amendment.

Far from being based on the commandments of the Bible, colonialism was the invention of the Canaanites who also invented chattel slavery. This slave-based culture was taken over by the pagan Greeks and Romans when the latter took possession of Phoenician mines and plantations. This same culture would eventually spread to the New World. If this pagan culture preferred by colonial European governments though strongly condemned in the Bible is not distinguished from the biblical culture of revivalist Christians who resisted and changed this culture, Christianity will appear the problem, being taught today. Fortunately, Christianity transcends and transforms while preserving the good and beautiful aspects of every culture.

Evangelicals have built home schools based on Creation Science. While Creationists have also developed a few journals and research centers, they have been unsuccessful in their attempts to introduce these teachings into America's public schools. One problem is that, though dating from the last half of the twentieth century, these schools are promoted and seen as outworn teachings from past. They lack the excitement of being connected to the recent discovery of Noah's Ark and unveiling of the profound truth that the supposed long Ice Age is due to having reimagined the deposits of the year-long Noah's Flood. However, these efforts have shown that Christians can develop and employ a standard scientific curriculum based on the Bible.

Part VIII

How the Discovery of Noah's Ark and Flood Will Change the World

68

WORLD CHANGING EVENTS

Gilbert Hovey Grosvenor was born in Constantinople Turkey, the Old World city dividing Europe from Asia that would be renamed Istanbul. As the second cousin of President William Howard Taft, Grosvenor was well connected to America's scientific and political establishment. He married the daughter of the great American inventor Alexander Graham Bell and was instrumental in developing America's National Parks system. Grosvenor understood how from the time of the 1925 Scopes trial, the scientific challenge to the Bible's account of origins had shifted America's trusted intellectual authority from the Bible to Science. This shift impacted the faith of Americans trained in the sciences because if one cannot trust the authority of the Bible regarding origins, how much less the inscrutable future! Long the world's most Christian nation and still sending missionaries to win the world for Christ, America's future was then being re-envisioned as leadership in industry, technology, and science! America would soon be sending men to walk on the moon, inspiring me to seek a career in science.

Instead of instructing visitors about God's *Creation,* Park rangers would refer to America's wildernesses as *Nature* which they explained in the light of the natural sciences. They had no idea that most of America's streams dated from the Flood or that the oldest native American artifacts date from before the Flood. As the case of most Americans, Grosvenor was not hostile to the Bible. He believed in the resurrection of Jesus though in Grosvenor's day most Protestants were being taught that miracles had ceased since the time of the Apostles. Christians such as Grosvenor who belonged to America's cultural elite believed that God had established scientific laws which scientists best understand. However doubtful of a possible discovery of Noah's Ark, Grosvenor knew that if found, it would not only shake the sciences but also world. It would be the greatest event since the resurrection of Christ.

How the resurrection of Jesus changed the world

The resurrection of Jesus marks the triumphant beginnings of the gospel. One cannot however appreciate the impact of the gospel apart from understanding the pre-Christian world including the classical civilizations of Greece and Rome. We tend to forget that the pre-Christian world was hardly secular. Far from a time of enlightenment, the world of the classical pagans was beset by the darkness of myths and by fear of capricious gods who had to be appeased through proper ritual. It was a time when some Europeans were sacrificing humans. But pagans everywhere regarded idols as being more sacred than humans. Christians however taught against worshipping worthless idols, things that God had created, and emperors who proclaimed themselves as a god. It was liberating to discover that the great and loving God of the Christians was sovereign over every power in heaven and earth.

Love was mostly absent in the harsh pre-Christian world especially beyond one's family and tribe. But Christianity was uniquely a religion of love and forgiveness. Jesus taught his disciples to love even their enemies. While the pagan world admired the wealthy and powerful, Christians showed concern for the poor and powerless including widows, orphans, strangers, and abandoned babies. Christians cared for the old and sick and especially slaves requiring excessive care that pagan families would abandon. Watching Christians being thrown to the lions for refusing to worship the Emperor, the pagans exclaimed "See how they love one another!" Christians showed no fear in the face of death. Eschewing worldly arms, the faith prevailed because Christians believed their Heavenly reward more valuable than their earthly lives. Living as a single Christian family, Christian communities thrived and spread throughout and even beyond the Roman Empire.

Pagan Greece and Rome could rule subject peoples only through public torture. For such purpose, the Greeks invented the terrible cross which the Romans would systematically employ to subject peoples who were not citizens of Rome. These most advanced of the pagan societies could prosper only through slavery. As I explained above, Plato and Aristotle taught the Greeks to be racist. While *classic* refers to a slave-based society, Christians regarded both free and slave as brothers and sisters. The faith would slowly eliminate slavery not just in Christendom but in most of the world.

Aside from what might be conceded to a stronger ruler or worthy opponent, forgiveness as necessary for peace and community was absent from pre-Christian peoples who were instead driven by a destructive desire for revenge. However advanced the arts and sciences of the pagan Chinese and regardless the great wisdom of Confucius, the pagan Chinese were continuously beset by warlords. If Buddhists monks could teach the wealthy how to obtain inner peace by eliminating self, they could not bring peace to a nation that would be ruled by the ruthless, take-no-prisoners Mongols. But apart from fear of the God of the Bible who avenges sin, changes human nature and values, ruthlessness will remain essential for social order.

The early Christians lived in their own communities as missionaries have continued to do, wealthy and poor living and working together as one Christian family. A contribution of the Christians to peace was their international community which unified the diverse cultures and nations where these communities existed and prospered. This is because Christians are citizens of a greater Sovereign than any on earth. Recognizing the advantages of working with this community, the Roman Emperor and the kings of barbarian nations voluntarily embraced the faith as had been prophesized in the Scriptures:

Nations will come to your light,
and kings to the brightness of your dawn. [Isaiah 60:3}

This is not to deny that the ways of the pre-Christian world continue to inform and influence Christians. As I will address below, Christians are beset by divisions. Fortunately, the born-again faith in which every believer discovers a personal relationship with the living Jesus can overcome these divisions. The Great Awakening revival of born-again Christianity united a dozen American sects and 13 separately governed English colonies as one Chrisian nation a full generation before 1776. The Continental Congress was subsequently formed from the intercolonial connections created by this revival of born-again faith. Revivalism continued to be the glue that melded immigrants from numerous countries and cultures into a single American nation. However the gospel teaches Christians to be good, productive, and loyal citizens and pray for the rulers of whatever nation or whatever place Christians happen to live. To the consternation of their enemies, the faith survives and thrives in places where exist the greatest hostility to the faith.

As people of the book, Christians pioneered universal literacy and learning which if existing in pagan cultures were limited to a few rulers, priests, and wealthy citizens. While pagan societies looked down on manual labor as the work of women and slaves, Christian men worked willingly and joyfully even in menial jobs. This led to the development of practical labor-saving technologies which were of little interest to the elites who ruled the slave-based societies, but which would create the Industrial Revolution in the world's most Bible-believing nations. Lacking partially to any faith, capitalists export these technologies to the world.

Aside from the wives of powerful rulers, wealthy prostitutes, and rare women warriors, women were hardly respected prior to the coming of Christianity which expects men to respect, protect, and provide for women, though not by requiring them to do things which only men can and should do. Christians pioneered charities and hospitals which until recent times have not existed among non-Christian peoples. Christian missionaries spread medical care, schools, and universities to non-Christian lands. It is thus appropriate that the Western world divides history from the birth of Christ.

THE DOLEFUL CONSEQUENCES OF THE UNANSWERED SCIENTIFIC CHALLENGE TO THE BIBLE

These great and positive changes to the world due to the Christian faith were forgotten as the men of the Enlightenment replaced a Christian/Heathen division of nations with white/colored races. As Science replaced the authority of the Bible, various kinds of racism became the secular world's scientific anthropologies. Unfortunately, the reaction to such racism by the children of Christian missionaries is responsible for creating the myths of the noble savage and that the peoples of the world once lived in peaceful harmony with one another and with Nature. So popular have become these absurd themes in revisionist histories, novels, and Hollywood movies that even Christians have come to accept this wholesale nonsense.

Though hardly a believing Christian, the philosopher Thomas Hobbes correctly observed the natural condition of mankind as being a life that is nasty, brutal, and short. A recent atheist, Tom Holland, explains how Christianity has resulted in the non-Christian world also becoming more humane.[62] To learn how Christianity continues to build nations, I refer you to my friend Vishal Mangalwadi's *The Book that Made Your World: How the Bible Created the Soul of Western Civilization*.[63]

---

[62] Tom Holland, *Dominion: how the Christian Revolution Remade the World* (2019, Basic Books)
[63] Vishal Mangalwadi, *The Book that Made Your World: How the Bible Created the Soul of Western Civilization* (2012, Thomas Nelson).

Vishal maintains the optimism about the growth of the Kingdom of God as promised in Jesus's parables and as preached in America's Second Great Awakening from the early days of the nineteenth century.

Unfortunately, nineteenth century liberal Christianity modified the faith to incorporate the scientific challenge to the Bible while still expecting a humanistic version of the Kingdom of God to spread progress, prosperity, and peace throughout the neocolonial world. These hopes were brought to a sudden end as nominally Christian nations fought one other in the racist-driven slaughter of the first World War. Reacting to the liberal and worldly version of the Kingdom of God embraced by the oldest and wealthiest mainstream churches, conservative Christians turned from expecting to win the world for Christ to looking for a future antichrist. This ended the traditional optimism of America, the first nation ever created by Christians specifically for the purposes of the Christian faith. It has seemed that the gates of Hell might prevail against Christ's Church unless the battle be somehow turned at the gates of the death of America and the West.

When the subjective foundations of modern science were discovered in the second half of the twentieth century, instead of bringing an end to trust in Science, truth was replaced by postmodern skepticism where "truth" is a matter of political and cultural power. Supposing there is more than one truth is a hypocritical way of denying truth, which is one and never changes. If seeming to make room for God and religion in a world which believes in Science, it merely creates competitors to truth and Christianity. We discover that traditional Christian beliefs based on the Bible are not to offend "progressive" choices. Such flexibility of truth also removes accountability from individuals and government leaders, putting the security and prosperity of a nation in great danger.

The result is that the people of America and the West become so divided that citizens can no longer work together to solve even their most dangerous problems which are ignored by the best Government that money can buy, and as private equity exploits and fires the high-priced talent responsible for creating technology and building industry. The benefits of Christianity disappear with nothing more consequential than dissolving the glue than once melded the diverse cultures and nations of America and the West.

Another unfortunate consequence of postmodernism is lack of interest in history and the greater matters of truth. But apart from the influence of the Bible, people lose interest in history. They also lose interest in larger issues such as the meaning of life for which scientific naturalism can provide only gloomy and depressing answers. To avoid pondering great problems and dangers, as also their own inevitable death, many learn to disrupt serious conversation with banter and attempts of humor while keeping themselves busy so as not to consider such unpleasant things. Such is life on the great "Titanic."

With the loss of moral standards, all manner of entertainment and worldly living, however decadent, can be purchased by those who can afford it. Those who cannot turn to alcohol and dangerous drugs even if they must steal to obtain them. Prosperity where it exists and government programs where it does not, together with the prideful celebration of a new sexual morality, only encourage dysfunctional families and with the latter, increasing unhappiness. The preferred solution to mental illness is drugs, which become the chief source of mental illness leading to murderous behavior. Whereas most Americans long expected to be more prosperous than their parents, few can now expect this. Not surprising, suicide has become one of the leading cause of death among lonely young and healthy people in America and the West.

A few serious young people seek identity and purpose, but due to the decline of historical understanding, today's universities can only produce justice seeking airheads whose unwise actions merely justify the very injustices they seek to end. This is because their professors have likewise learned their history from fashionable "historical" novels, Hollywood movies, revisionist histories, and from teachers and colleagues influenced by the same. Nothing is more dangerous than the "righteous" mind, by which is meant a judgmental mind that leads not only to violence but to intolerance and totalitarianism. Such has been the outcome of all ideological utopias since the French Revolution.

Non-Christians who still think recognize the problems created by a post-Christian world. A few non-Christians thinkers such as Jordon Peterson, Tom Holland, and even the famous atheist Richard Dawkins who was formerly hostile to Christianity have begun to recognize the importance of Christianity to society. But the problem cannot be solved by returning to traditional liberalism or cultural conservatism. These thinkers have long faced an intellectual barrier which prevents them from embracing the faith, namely, the scientific challenge to the Bible! They may not be shocked to learn that the narrative of the past taught by the current historical sciences has blinded us to the truth, but they surely will be shocked when they discover the truth is the simple narrative of origins put forth in the Bible.

My purpose here is not to remind you of what you already know. Other than mentioning the root cause of all these problems, my aim is to raise a banner of hope for those who find their identity as part of the God's family created by the gospel. We have God's promise that

In that day the Lord Almighty will be a glorious crown, a beautiful wreath for the remnant of his people. He will be a spirit of justice to the one who sits in judgment, a source of strength to those who turn back the battle at the gate. [Isaiah 28:5-6]

69

How Naturalistic Science Beguiled Christians

How did a Christian nation became so enveloped in darkness? Other religions may also be based on a supposed salvation, a way of life, a worldview, or certain religious practices or confessions but Christianity is uniquely based on hearing, holding to, and teaching truth. Otherwise, the Christian faith cannot exist, cannot even be understood. If however we come to believe that our faith is not based on truth, we should cease being Christian. This is because authentic Christianity is a way of suffering and service. The Apostle Paul reminds us that if Christ be not resurrected from the dead, we are of all people the most to be pitted.[64]

However, due to being based on historical truth, Christianity appears unreasonably vulnerable to refutation by historical or scientific investigation. The easiest thing in the Bible to test and perhaps refute was the account of Noah's Flood, the evidence of which naturalistic science managed to explain away, leading to the supposed antiquity and evolution of man with devasting effects to the Christian faith.

Science was able to beguile Christians because it also purported to be based on truth, leading to the supposition that born-again Christianity is based on the imagination and the Scriptures on fabricated testimony. Truth is coherent and thus cannot conflict with itself. On this basis, the colleges of America founded and led by Christians opened their doors to scientific teachings which, in the beginnings of modern science, were mostly the product of Christian ministers and supposed as being at least in conformity with the

---

[64] 1Corinthians 15:17-19

teachings of the Bible. However subsequent scientific teachings would oppose Christianity even in the social and moral arena.

For example, Yale University President Noah Porter (1811-1892) was concerned when libertarian William Graham Sumner (1820-1910) planned to use the agnostic Herbert Spencer's text on sociology.[65] The originator of the phrase 'survival of the fittest' and what would become known as *Social Darwinism,* Herbert Spencer (1820-1903) compared Christianity to the superstitions of South Sea islanders. Sumner was himself famous for teaching his 'root hog or die' libertarian ideology. When their dispute become public, Sumner won the debate on the basis that education at Yale ought to be based on truth whatever be the source. Sumner himself abandoned Christianity for Social Darwinism, the very opposite of Christianity. Though ignored by America's Supreme Court, Porter pointed out that atheism and agnosticism are as much religious creeds as are theism and dogmatic Christianity. Naturalistic science is not only a faith-based religion but has proven to be a truly evil religion.

This is how supernatural-denying naturalistic science undermined the Christian foundations of American colleges. Established for the purpose of teaching the truths of Christianity, America's colleges were turned into secular institutions teaching young Christians to reject and attack rather than defend the biblical faith. This immediately led to the new scientific universities such as Cornell, John Hopkins, the University of Chicago, Carnegie Institute, and Stanford that were sponsored by monopolistic industrialists who aimed to see Social Darwinism taught in America's leading schools. This financial, industrial, political complex continues to oppose the morality of the Bible less it restrict pleasure, corporate profits, and markets.

Academic freedom and tenure were soon introduced into America's colleges in order that things opposed to biblical Christianity could be freely taught. Academic tenure had begun in Germany for the same reason as in America, to allow professors to teach without hindrance things contrary to the Bible and Christianity! This happened as Science was challenging the Bible regarding the origins of life and mankind.

Whereas a society based on voluntary Christianity has little problem allowing all manner of contrary beliefs, societies not built on voluntary Christianity will see a totalitarian state whether of the political right or

---

[65] c.f. George M. Marsden, *The Soul of the American University: From Protestant Establishment to Established Nonbelief* (1994, Oxford University Press)

left as being necessary to protect its teachings and practices. Many things prohibited by the teachings of the Bible may appear a nuisance or a threat to a morally libertarian society. Thus, a truly free society can exist only if grounded in biblical Christian faith!

One thing responsible for the deception of science is the supposition that science is based on "verifiable" evidence, a teaching that dates from Francis Bacon. However, the great philosopher of science Karl Popper (1902-1994) pointed out, even the most contradictory evidence can be spun to *verify* any scientific theory. He used examples from Sigmund Freud and Felix Adler's competing theories of psychoanalysis and also from the Marxists' dialectic materialism to show how these theories could come up with *ad hoc* explanations to turn the most contrary evidence into examples of verification. Popper however contrasted what he regarded as these pseudo-scientific theories to Einstein's theory of General Relativity. Einstein had pointed to a testable but otherwise unexpected observation of star light being bent by the sun's gravity and potentially disproving his theory of General Relativity. Thus, Popper asserted that instead of verifiability, science should be based on *falsifiability.*

This is to say that truth should make specific and non-trivial predictions so as to be *testable*. To elaborate on Popper's insight, a non-falsifiable theory as Darwin's theory of evolution by natural selection can provide no guidance for discovery. Neither can the theory of man-made climate change, though both theories can explain everything that has or ever can happen. All that is needed is a movement to marginalized those who question these theories, together with economic or political rewards for those who promote them.

In contrast to theories such as evolution by natural selection and man-made climate change, the Bible is lavishly falsifiable due to being grounded in historical people and events. However, for the same reason it is lavishly enlightening. This is why the historical narratives in the Bible provide a powerful light for understanding the world and its history.

70

How this Discovery Will Change the World

This discovery is far more than just the discovery of Noah's Ark. It is also the discovery of Noah's Flood, the discovery of the antediluvian world, and the discovery of the literal truth of the Bible while it reveals the thick darkness that has blinded our modern world to the truth concerning origins.

1 - It confirms the Bible's most unbelievable miracle

We now have powerful material evidence for what is arguably the greatest Nature miracle in the Bible, the most difficult miracle in the Bible for a scientifically trained person like me to believe. The discovery is a death blow, a public humiliation of scientific modernism as well as supposed 'scientific histories' which reject reports of miracles. This discovery not only confounds unbelieving modernists but also Christian teachers who have tried to accommodate faith to science either by twisting the Scriptures to harmonize them with the current science or else by assigning faith to a separate realm from the one studied by science,

Fig 55 – Doubting Philip touches one of the wooden nails that held together Noah's Ark

thereby emptying the faith of its supernatural power. Though teaching the visible Creation as coming from the invisible spiritual world of Heaven, the Bible also addresses the creation and history of the physical world. The light of the Bible is thus necessary for understanding the historical sciences. The honest response for doubters like me is to admit to having been deceived. There is nothing more liberating! I was deceived by my scientific education, but it has taken more than forty years to discover exactly how I was deceived. I was doubting Philip until my quest which ended when my fingers touched one of the wooden nails that held together Noah's Ark.

However important the revelation of the truth of this great miracle, little is gained simply by establishing the reality of miracles. While miracles do refute naturalistic science, times have changed since Grosvenor shared his predictions of the consequences of discovering Noah's Ark. There has probably never been a time in the history of America when more people testify of having experienced a miracle. This is mostly due to the Pentecostal and Charismatic forms of Christianity and especially reports from the non-Westernized nations where this form of Christianity has driven the growth of Christianity. Unfortunately, it is also due to New Age forms of Christianity which reject science without embracing biblical Christianity. Too many supposed miracles but cheapen the coinage of miracle without contributing to biblical faith! I am not supposing that many have not witnessed or experienced authentic miracles. I have myself. Rather, I am pointing to the fact that none of these have the same world changing effect as the public miracles of the Bible. But we do now have good evidence of one of the greatest world-changing miracles mentioned in the Bible.

2 – IT WILL BRING AN END TO THE THEORY OF MAN-MADE CLIMATE CHANGE!

The theory of man-made climate change is based on what have been wrongly supposed as past Ice Ages when glaciers covered much of the earth. The earth did become warmer following the Flood that occurred less than five thousand years ago, but most if not all of the polar ice caps date from that time.

Even if we have been taught to ignore God's blessings and judgments, our invisible good God has always governed the physical world including changes in weather and climate. Has not the Lord warmed the earth specifically to add carbon dioxide so as to make agriculture more productive, thereby providing for the earth's greater population since the Industrial Revolution!

What arrogance that man should now suppose he has mastery of the earth's climate, which he did not create! Recognizing that the Ice Age was based on re-interpreting the evidence of Noah's Flood will put an end to the myth of man-made climate change that has resulted energy policies that have a great and detrimental effect on the poor, who must spend a far greater part of their small incomes on energy. This problem cannot be addressed by government programs supposed to benefit the poor but which only subsidize irresponsible behavior and the businesses that exploit these programs. Giving subsidies to junk old cars benefitted automobile manufacturers but kicked the poor in the stomach by increasing their cost just for driving a car. Those who make such laws may desire for the poor to eat cake, but they are oblivious to the fact that the one must pay subsidies just for being or even living with the poor. By abandoning the gospel, America's leaders and people have destroyed the world's first middle-class nation.

3 – It establishes the Scriptures as the foundation of truth

The most profound importance of the discovery is its witness to the truth of the early chapters of Genesis as it undermines the so-called higher criticism of the Bible. Archaeology has always been the nemesis of the higher criticism of the Bible which dates from before the beginnings of archaeology. The publication of the Dead Sea Scrolls back in the nineties put an end to the old liberal criticism of the New Testament by confirming the New Testament's account of Second Temple Judaism including the fact that the Old Testament text used at the time of Jesus is essentially the same as that of our present Bibles. A plethora of finds have confirmed the biblical history of ancient Israel from the time of Abraham. That such archaeological discoveries can be understood from the narratives in the Bible can only be the case if these narratives accurately reflect those ancient places and times. This points to the authentic historiography of the narratives of the Bible. Though there remains confusion concerning such things as the Exodus due to not recognizing the Flood, no archaeological discovery has ever contradicted the Bible.

However, discoveries related to the early chapters of Genesis which include the Flood of Noah lie at the foundation of the JEPD composition criticism which, as I explained in Chapter 2, supposes that the book of Genesis was composed no earlier than the Babylonian captivity in the seventh or eighth centuries before Christ. The presupposition of JEPD is that the Ancient Near Eastern accounts of the Flood such as the Gilgamesh epic predate the account in the Bible, which is claimed to have been borrowed from these Ancient Near Eastern texts. These accounts do not however predate the era of the Flood as given in the Bible. There is no reason to doubt that they are independent though inferior and confused accounts of the same event. Because the recent discovery is so in line with the account of the Bible including the landing place of the Ark settles the question of which account is more accurate. It demolishes the supposition of the author of Genesis as having borrowed from the Ancient Near Eastern texts. Many scholars taught these things by scolding their students from the authority of their academic positions. This discovery puts an end to the arrogant boasting of "critical" biblical scholarship.

Such an accurate account as preserved in the Bible could hardly have been the case aside from written records which we may suppose were possessed by the Hebrews who descend from Shem. Many have supposed that this account was given directly to Moses by divine revelation, the view of the Jewish rabbis. Such a supposition however is not in line with God having used written texts to preserve his testimony. Such a view of this early history of the world first being revealed to Moses raises the question of whether the patriarchs Abraham, Issac, and Jacob possessed the knowledge of God's Creation of the world, that all mankind descend from Adam and Eve, and what for them was a not so long ago Flood and the even more recent scattering the nations from Babel. Moses reminds Israel that their fathers remember this latter event.[66]

The Genesis account of the Flood reads much like a ship log with a calendar based on the birth of Noah, the same calendar used by the ancient Chinese who descend from Shem. We know that writing existed in precisely this part of the world even before the time of the Flood. According to the famous Sumerian scholar Samuel Noah Kramer (1897-1990), the name *Sumer* derives from Noah's son Shem. Other philologists tell us that the biblical land of *Shina*r is a reference to the lands of the kings of Summer and Akkad. As I mentioned above, the tree of life motif is engraved on some of the pottery that we found on the Ark.

The simplest explanation for the composition of Genesis is that God inspired Moses to write it from the records preserved by the Israelites living in Egypt using a Hebrew alphabet created by the same Israelites living in Egypt. Though the Hebrew language was borrowed from the people of Canaan, the Canaanites preserved their records, including the Gilgamesh account of the Flood as found in the ruins of Megiddo, in

---

[66] Deuteronomy 32:7-8

cuneiform! Thus, it is more likely that the *Phoenicians,* as the Greeks called the Canaanites, borrowed their alphabet from the Israelites who lived in this land from the time of Joshua and dominated the land from the time of David and Solomon. Whatever editing occurred to update the names of places, the history of Genesis is in line with history as preserved by Israel's Levites, priests, and kings from the time of Moses. These are the Scriptures and history affirmed by the same Jesus Christ who God used to create the world and instruct Noah to build the Ark.

This recent discovery of Noah's Ark and this true understanding of the Flood which allow us to understand the discovery does not answer every question pertaining to the Bible. It does however establish the Bible as the presumed authority regarding every question pertaining to origins and mankind's ancient history, something that has long been presumed the province of naturalistic science and various historiographies revised to accord with naturalistic science. Because history includes contingent acts including miracles, a supposed 'scientific history' is but a foolish oxymoron.

It is also true that some have remained faithful to the Scriptures regardless the claims of naturalistic scientists and academic scholars. Otherwise, there would be no Christian faith. There would also have been no Bible, which has been preserved, not by any institution, but by Christ's true church, the pillar and foundation of truth.

Regardless of what some claim and however important and valuable textual scholarship, only those filled with the Holy Spirit can identify the writings belonging to the inspired canon and these alone can understand the meaning of the inspired Scriptures. The Holy Spirit who gives understanding to the Scriptures is the result of believing the narratives of the Bible concerning Jesus. Understanding the Scriptures is therefore outside the province of mere academic scholars whose opinions should mean little to those whose beliefs this discovery has vindicated. This however is no excuse for pride or arrogance on the part of believers, which is something that God hates. Believers also depend on God's mercy. Moreover, other that appealing to faith, we believers have not previously possessed those answers, regardless of what some have pridefully claimed. Rather, believing will always require humility and repentance. However, we will now be able to defend the Bible's account of origins not only with our hearts but also with our minds!

4 – It exposes the "Enlightenment" as having been a sheet of thick darkness covering the minds and hearts of our Modern World

To fully understand the impact of this discovery, bear in mind that what is known as *the Western world*, or simply *the West,* is based on (1) the rejection of miracles, (2) the rejection of divine revelation, and (3) dismissing the unseen spiritual world. This began with the Enlightenment which developed from rationalism and Isaac Newton's materialistic science. The importance of Darwin's theory of evolution by natural selection, which I have scarcely addressed in this book, was its supposed intellectual support of atheism. Coming to be regarded as accepted science, Darwin's theory of evolution licensed atheism. Because Darwin's theory of the evolution of man rests on the antiquity of man, it is collateral damage of this demonstration that the evidence of the Flood was wrongly interpreted as a long Ice Age.

71

How the Scientific Deception Freed Faith from Government Control

Not everything about the scientific deception has been detrimental to the faith. It has also resulted in the gradual freeing of Christianity from control by worldly governments and the freeing the body of Christ from the kind of institutional Christianity that has been used by governments to control faith. This began with America's separation of the church and state, never intended to protect the state from Christianity, but rather the reverse.

The New Testament teaches believers, who are those called out from the world, to keep themselves separate from the world which is under the authority of worldly governments! However believers honor earthly governments while serving as ambassadors of the coming Kingdom of God. While some this-world governments became nominally Christian, such "Christian" governments would naturally use the faith to enhance and exercise their worldly power, thereby corrupting the faith. Resisting such control subjected Christians to persecution which was often more severe than from non-Christian powers. While God appoints rulers of worldly governments for his good purposes, the kingdoms of the present world are not of the coming Kingdom of God. Only those who obey the living Jesus and follow his Spirit are functioning members of the Kingdom of God. The governments of this world will however eventually come to the light of God's separated people.

America's new secular order easily accommodated the separation of church and state but was also influenced by the Enlightenment from which developed the scientific challenge to Christianity and the Bible. This separation of church and state began with mutual respect between the new Government and Christianity. This continued to be the case so long as America's colleges and schools were sponsored and run by the churches and so long as the churches resisted the deception of the Deistic Enlightenment. However, I have shown how the deception of Deism and the Enlightenment infiltrated the church-operated universities and churches beginning with the state-controlled churches of Europe. This however diminished the influence and growth of the state-controlled churches. Christianity's growth shifted to the separatist born-again Christianity centered in America. Christianity has since been returning to its independence of state control as existed in the beginning of the faith.

The faith of course now competes with state sponsored and controlled schools which teach things opposed to the faith. Those of faith may also suffer from laws and regulations that threaten religious freedom. However, after discovering the truth concerning Noah's Ark and Flood and how simply the origins of the world can be understood in the light of the Bible, those who love truth will desire that such historical truth be taught in the public schools, at least as an alternative to the current scientific understanding. Otherwise, Christians will prefer our own schools. Those who attend schools teaching truth will become the leaders of the next generation.

Many have unnecessarily feared that returning to a Christian nation means a return to a theocracy, by which is meant a worldly government headed by a religious clergy. Neither do Christians wish to see their secular governments headed by some clergy as some Christian nationalists may wrongly teach. To the contrary, as was also the case at the beginning of Christianity and to a great extent the early days of the American Republic, this means spiritual, moral, and intellectual light provided by **the example** of a flourishing Christian community which is independent though respectful of secular governments.

The Christian faith does not depend on the worldly power of governments but on the greater spiritual power that comes through the weakness of the cross. The cross means that Christianity must always remain a voluntary faith led by those willing take up their cross and follow Jesus. Thus, biblical Christianity flourished far more in America than in the nations of Europe whose governments supported, or in some cases as in Great Britain and Germany, headed their official churches. Though separated from the state, Christians should not abandon education, medical care, or charity to the government. Rather, they should provide these services to their own communities as was the case in early America, a tradition which partly continues in America's Bible Belt. Secular governments have enough to do without doing what they do badly.

A secular government as we have in America aims neither to patronize nor proscribe any particular religion. Notwithstanding such government neutrality, we hope the aim of public schools and those who support them is to impart useful truth. If so, the students will benefit. If not, the government and the society to which these students belong will suffer from not being taught that which is true and instead being taught that which is not true.

Attending public school in America during the fifties and early sixties, my generation were taught history from the Bible. This was ended by a Supreme Court ruling during my high school years. As it turned out, only a theory of origins contrary to the Bible could be taught in the public schools. Instead of protecting religion from the government as the founders intended, the Supreme Court effectively implemented the official religion of scientific naturalism which is opposed to Christianity. That American courts have refused to allow competitors to scientific naturalism be taught in America's public schools is proof that scientific naturalism was in truth the official religion of America's Federal Government!

The scientific deception which resulted in the removal of religion from control by governments in Europe and America may however have been God's plan to free the body of Christ from control by secular governments by sending tis powerful deception called Science![67] Though Christians must contend with attacks on their faith from a post-Christian culture that has as its patron all the resources of the post-Christian government and culture, truth is more powerful. It will free those willing to pay the price to follow Christ. This same passage tells us that the darkness of those who refuse to acknowledge the truth will be destroyed by the light of Jesus' Coming rule.

---

[67] c.f. deceptions mentioned in 2Thessalonians 2:11 and warnings from Jesus concerning last day deceptions from sorcery and false signs. These would include the marvels of science. Such deception by those who despise truth will be destroyed by the light of Jesus' Coming.

72

Why the Scriptures Must Become Believers Essential Authority for Discovering and Testing Truth

Apart from some established authority, we have no way of judging what claims might be in line with truth. Thus, from the beginning of the faith, Christians recognized the authority of the Bible, the foundation of the faith once and for all time entrusted to God's holy people.[68] Sadly, even many Christian teachers have lamented the “anti-intellectualism” of trusting biblical authority. There has never been a greater American mind than Charles Sanders Peirce (1839-1914) the famous American philosopher and historian of logic. Peirce explained that even in the case of science, authority is essential. Thus, what the men of the Enlightenment called “free thinkers” leading to what would be called liberalism can never lead to truth. A liberal or enlightened man is one whose mind has been freed from religious authority though not from darkness.

Beginning with Gottfried Ephraim Lessing (1729-1781), Enlightenment thinkers demanded that the sacred Bible be examined for truth as might any profane book. But what did the Enlightenment thinkers use for testing the truth of the Bible? Lessing rejected any contingent or "accidental" events of history as the basis of rational truth, which had instead to be based on the 'universal and necessary truths of religion.' By the latter, Lessing meant something like Spinoza's pantheism. Even had the Bible not reported miracles, the historical narratives by which the teachings of Jesus are recorded in the Bible could not have served as Lessing's standard of truth. Unfortunately, this continues to be the way that supposed 'critical scholars' judge the

---

[68] Jude 3

probability that various sayings of Jesus can be suppose as historical. It was this same bias that resulted in the rejection of the Bible as a light for science and also for rejecting contingent catastrophes such as Noah's Flood to explain the geological record. On the other hand, these "critical" scholars could accept almost anything that could explain the Bible's composition in some non-miraculous manner. Albert Schweitzer (1875-1965) showed how the Jesus of history discovered by various critical scholars always conforms to the worldview of the critical scholar.

Thus, did the thinkers of the Enlightenment reject all the miracles of the Bible on the basis that men of the Enlightenment reject miracles! They were their own shallow but prideful standard of truth. Their standard of truth was based on what was or could be believed by the Enlightenment's rationalist thinkers. What various specialized scientific communities accept as true continues to be the standard for testing truth in peer-reviewed science. But how do those who are not part of the community of specialists know that these scientific communities are any different from the "critical" thinkers of the Enlightenment?

We can use Jesus's criterion that we can know truth by its fruits. This is a valid way of testing the science used in practical arts and inventions. However, we must not suppose that speculations regarding origins using scientific principles have the same validity unless also showing the same good fruits. Rather, it can lead to a great deception.

This however is little different from how various communities of Christians determine truth, though God may not have inspired the leaders and teachers of these Christian sects in the same infallible way as in the case of the Apostles and Prophets of the Bible. I will address this problem in the next chapter.

Because some miracles are real, science can never serve as an authority for testing historical claims. Indeed, God has and may continue to occasionally change the scientific order of the world as he did on each successive day of Creation and at the time of the Flood. Many have declared that such changes in the order of the world makes science impossible. To the contrary, this is a help rather than a problem for science because these world-changing miracles are recorded in the Bible! Scientists must however humble themselves and examine their evidence in the light of the Bible.

It remains true as in the case of all words, texts, and scientific observations, that the words of the Bible must be interpreted. Fortunately, we do have an interpreter of the Scriptures, the words and

Spirit of the living Jesus, which is also the spirit of truth. Thus, we should interpret the Bible through the words of Jesus recorded in the Bible, understand the words of the Apostles in the light of Jesus's words, and understand the words of the Old Testament Prophets in the light of the New Testament. Such is the way of the Church that is built on the foundation of the Apostles and Prophets with Jesus Christ as the chief cornerstone.[69] It is the way that I tried to read the Bible so as to use its light to examine the evidence from the ancient world.

In the process of this investigation, I discovered that the more light from the Bible that I used, the more light I obtained for understanding the evidence of the ancient world. And the more light I obtained for understanding the evidence of the ancient world, the more light I obtained for more completely understanding the evidence of the ancient world. In the beginning of my search for evidence of Noah's Flood, I was surprised to see this happening. As I discovered the words of the Bible proving themselves time and again, I repented of my doubts and accordingly gained all the more confidence not only in the words of the Bible, but also the results of using this light for the purposes of science and history.

[69] Ephesians 2:20-21

73

How this Discovery Can Help solve Christian Divisions

The main subject of this book concerns the truth about origins. As great as been the problem of the scientific challenge to the Bible, an equally great challenge to the faith are the divisions among Christians whether caused by competing ancient churches, sects, or divisions in Christian teachings and practices that transcend the traditional divisions. However, should Christians come to consensus about the most important matters of the Christian faith, the church of Jesus Christ will become a great light to the world.

This is not to be naive about the divisions or the possibility that the various traditional churches and sects can or should ever be united. But regardless these divisions and transcending these earthly divisions, we know there is one body of Christ consisting of all who have a direct relationship with the living Jesus. Trusting in Jesus Christ and the Bible that he inspired are the two essential things about which all Christians agree regardless our differences concerning non-essential things and however important the latter. Due to their divine origin, all Christians should hold these two truths – the living Jesus and the inspired Scriptures – and only these two truths as divine dogma. But if one rejects salvation as coming exclusively through Jesus Christ or rejects the Bible as the foundation of the Christian faith, he or she should not be regarded as a Christian brother or sister in our common faith. Of course, we should seek to win or restore such ones to the true faith.

Jesus tells us that the way to life is narrow, and few are they who find it.[70] As was also the case of ancient Israel, there have been times in

---

[70] Matthew 7:13-14, Luke 13:23-24

which those holding to the truth have been few, which may also be the case today. Who but God can and should judge hearts and minds?

As happens in all revivals of the faith, embracing truth begins with individuals. Though organizations and institutions have their good purposes, those who lead and teach within institutions have an obligation to the organization to which they belong be they ecclesiastical institutions, schools, or ministries. Unless one has the cooperation of their organization, he or she may lack the freedom to pursue wherever truth may lead apart from paying whatever sacrifice may be required to pursue truth. Truth will make you free, but as I will discuss in Chapter 77, some must pay a price for truth to prevail.

Our faith will be most effective when the members of the body of Christ demonstrate our ability to resolve our differences so as to believe and teach the same thing.[71] This of course is a matter that only Jesus Christ can solve, though he has promised to do so at the end of the age when he separates the tares from the wheat. This parable of Jesus explains the division in the Christian faith which have existed since the time of the Apostles.

"...The kingdom of heaven is like a man who sowed good seed in his field. But while everyone was sleeping, his enemy came and sowed weeds among the wheat and went away. When the wheat sprouted and formed heads, then the weeds also appeared.

"The owner's servants came to him and said, 'Sir, didn't you sow good seed in your field? Where then did the weeds come from?'

"'An enemy did this,' he replied.

"The servants asked him, 'Do you want us to go and pull them up?'

"'No,' he answered, 'because while you are pulling the weeds, you may uproot the wheat with them. Let both grow together until the harvest. At that time I will tell the harvesters: First collect the weeds and tie them in bundles to be burned; then gather the wheat and bring it into my barn." [Matthew 13:24-30]

Then he left the crowd and went into the house. His disciples came to him and said, "Explain to us the parable of the weeds in the field."

He answered, "The one who sowed the good seed is the Son of Man. The field is the world, and the good seed stands for the people of the kingdom. The

---

[71] 1Corinthians 1:10

weeds are the people of the evil one, and the enemy who sows them is the devil. The harvest is the end of the age, and the harvesters are angels.

"As the weeds are pulled up and burned in the fire, so it will be at the end of the age. The Son of Man will send out his angels, and they will weed out of his kingdom everything that causes sin and all who do evil. They will throw them into the blazing furnace, where there will be weeping and gnashing of teeth. Then the righteous will shine like the sun in the kingdom of their Father. Whoever has ears, let them hear. [Matthew 13:36-43]

We may be seeing the beginning of this promised gathering. The shift in America and the West to a post-Christian world which began with the French Revolution, and which has now reached America has caused Christians of different persuasions to appreciate one another as we contend together for the truth of the Bible.

Such a common understanding can begin by coming to a consensus of what the Bible teaches about origins, beginning with the matter of Noah's Ark and Flood. Fortunately, believers from the spectrum of Christianity have been involved with this Noah's Ark project, from Pentecostals in Asia and Evangelicals in America to Roman Catholics in Europe. There is, of course, only one Noah's Ark. This creates an opportunity as is already happening for growing unity between all Christians. This discovery is certainly going to impact the world of Christianity if not the whole world.

Even if we discover the truth, we cannot be arrogant. Rather, as the Apostle Paul tells us, we must

Be completely humble and gentle; be patient, bearing with one another in love. Make every effort to keep the unity of the Spirit through the bond of peace. There is one body and one Spirit, just as you were called to one hope when you were called; one Lord, one faith, one baptism; one God and Father of all, who is over all and through all and in all. ... until we all reach unity in the faith and in the knowledge of the Son of God and become mature, attaining to the whole measure of the fullness of Christ. Then we will no longer be infants, tossed back and forth by the waves, and blown here and there by every wind of teaching and by the cunning and craftiness of people in their deceitful scheming. Instead, speaking the truth in love, we will grow to become in every respect the mature body of him who is the head, that is, Christ. [Ephesians 4:2-6.13-15

74

What Christians can Learn from Scientists

In regard to discovering truth, there is something that believers can humbly learn from scientists and historians. I once heard a philosophy professor explain the difference between science and religion. While science holds knowledge provisionally, religion supposes that it already possesses the truth. The advancement of science entails that science does not already possess complete knowledge. The same has always been true of Christians who grow in faith and maturity. But the professor was telling the truth in the case of institutionalized religion, those who the theologian Karl Barth sarcastically referred to as 'the blessed possessors of the faith.'

Christians do of course possess the Scriptures which never change, the once and for all time foundation of the faith. This church of the living God has always stood as the pillar and ground of truth.[72] Unfortunately, even Christians like me who ascribe to *sola scriptura* disagree about what the Scriptures teach. Christians who assign the task of interpreting the Scriptures to some ministry other than Jesus Christ will be disappointed. Religious pride, which God hates, keeps us from seeking to resolve our differences.

It is also wrong to suppose that we can understand the Scriptures without the wisdom and correction of other believers. Thus, another thing that believers can learn from science is the importance of building on the work of those who have preceded us. We do not realize how much we already read the Scriptures in the light of past theologians. Isaac Newton was not the only great scientist to have claimed to be able to see further by standing on the soldiers of giants. Newton had in mind those such as Copernicus, Kepler, and Galileo. Likewise, Einstein built

---

[72] 1Timothy 3:15

on the work of Newton. The same is true in every field of science. It is also true in the development of technology and industry. The Scientific Revolution occurred during a time when there existed a common education in the humanities. Scientists build on the knowledge of the past as well the work of their contemporaries.

This is even more the case with historians, including historians of ideas from whom I have learned much. Far more than in the case of science, historians tend to come to a consensus. Understanding the historical development of philosophical and theological teachings is key to a deeper understanding of philosophy and theology. The same is true of technology. I made my contributions in the field of telecommunications because I understood the history of telecommunications technology. Steve Jobs used his knowledge of the history of printing and computer interfaces to introduce truly personal computing.

Much can be learned from the historical circumstances that gave rise to certain philosophical and theological teachings. For example, in the earliest days of the faith, a monarchical and hierarchical episcopacy helped save the Church by protecting the Scriptures from the Gnostic gospels. Though it was Jesus Christ and not the Catholic Church who gave us the Scriptures, God did use the Catholic Church to preserve the Scriptures. Unfortunately, by the time of the Reformation, emperors and kings were using the episcopacy to control the churches and prevent believers from reading the Scriptures. A view of the sovereignty of God which denied freewill helped deliver believers from powerful political and episcopal control. However misapplied forms of these teachings are being used today to excuse fleshly and worldly behavior which should not be the case of God's holy people. When teachings are part of our religious tradition or when we have made the defense of certain teachings part of our identity, we might not be inclined to listen to those who challenge our tradition. Our identity ought instead to be in Jesus Christ alone.

Fortunately, conflicting views concerning origins are more recent and thus most are more willing to see their own views examined and criticized. This makes the truth about origins an ideal place in which to make progress towards a consensus of all views shared by Christians.

75

THE JOYS AND CHALLENGES OF PURSUING WORLD-CHANGING TRUTH

I have been explaining the recently discovered Noah's Ark and the misunderstood evidence of Noah's Flood as if these are but scientific and historical matters. It is good to be detached when one is seeking an objective understanding, but this has hardly been an ordinary investigation. There have been occasions of elation when I obtained a new insight or when I embarked on a memorial journey to learn more about the archaeological evidence of Noah's Flood and this recent discovery. Not only did I visit the second deck of Noah's Ark, but I was honored to be the first outsider to meet the family who had kept secret the Ark's location and artifacts.

I had never even dreamed of doing any of this. From the beginning of my search, my aim had been to understand the greatest scientific challenge to the truth of the Bible. As I was about to publish the first ever understanding of the archaeological evidence of Noah's Flood, I heard my mother who was in another room yelling for me to turn on ABC Evening News. Diane Sawyer was announcing this discovery, fulfilling a prophecy from 10 years previously, which I could not then believe — 'When you complete your book, Noah's Ark will be found!' Imagine my joy when I heard the announcement!

There have also been great challenges. The very next evening, Dianne Sawyer announced that the discovery which she had announced the previous evening might be a hoax. Though I knew that my work would be essential for understanding the discovery, I did not expect to at once become its chief defender. Only a few week after their world-changing announcement, *NAMI's* Clara Wei contacted me to visit my city and speak to a private group which my friends and I had gathered.

Being neither a field archaeologist nor a mountain climber, I never thought that I should be one to visit the site. However, it soon became clear to be that this would be essential to my defense of the discovery, even by some who knew me but not the discoverers. Never once did I ask to visit the Ark. Parasut decided that he needed to get me on the mountain to reawake interest in the discovery which was burdened by those charges of hoax. When I received Parasut's call, my mother with whom I lived was in the late stage of hospice. She and my sister who was assisting with her care strongly insisted that I go to the mountain.

As we arrived at the airport near the mountain, we discovered that we were being followed by a menacing team of spies. They had learned of our expedition from my online journal. It took a full week for the spies to tire of following our advanced team who were already on the mountain. After the spies gave up and came down the mountain, a deadly storm destroyed our mess tent causing our team to have to run down the mountain just to keep from freezing to death. We lost our professional videographers who we were expecting to document our climb. It took another week to manufacture a new and stronger tent.

As we waited in the hotel, I called Sheila each morning to remind her to pack her gear to climb until the very morning that I gave up expecting the call. Needing to return to my mother's side, I resigned myself to the dismal probability that we would return home without even climbing the mountain. I did not look forward to informing my friends and supporters, especially my dying mother.

Fig 56 - Looking over Turkey and Iran from about 14,000 feet during the most miserable night of my life

That is when Parasut called to ask where we were staying? Our hotel happened to be on the north side of the mountain at an altitude of 2,800 feet. In 6 more hours, we were at 14,000 feet where Sheila and I spent the most miserable night of our lives. The astronaut James Irwin said that it is more difficult to walk on Mount Ararat than walk on the moon. I believe that sleeping on Mount Ararat is more miserable than sleeping on the moon, especially when one has to pack in a great hurry.

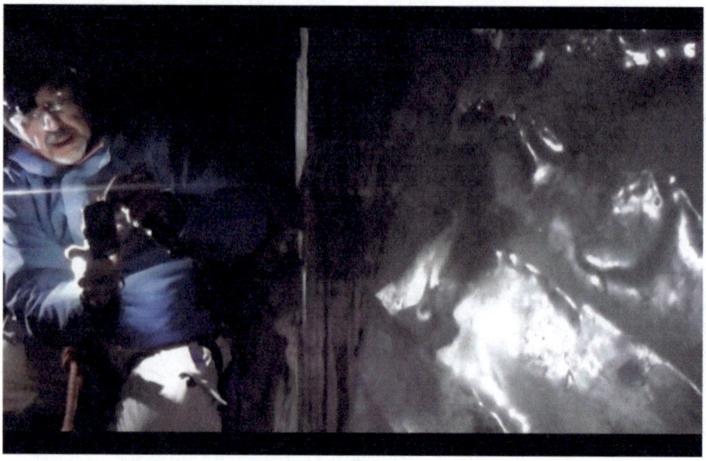

Fig 57 - Photographing the bowed walls of Noah's Ark

When we arrived at the passage to climb inside the Ark, I was too sick to enter! Having come so far and endured so much, I could only sit and rest. While Shiela was becoming the first American to visit Noah's Ark, I miraculously recovered from my altitude sickness to get my own wide-eyed tour of this amazing structure which I recorded with my borrowed camera and iPhone.

After exiting the Ark, my altitude sickness returned. Having packed high camp, we had to get off the mountain before nightfall. I could only slowly walk a few feet before I had to stop and rest. Parasut hurried Sheila down the mountain, leaving Davut Gimril, his second in command, with me. I expected to die on the mountain, but I was consoled that Sheila would return to tell about visiting what I knew can only be Noah's Ark.

I was in a delirious state when Davut shouted that Parasut had sent a horse up the mountain to rescue me. We reached base camp just as the sun was going down. I barely remember Davut standing on the running board to keep our Land Rover with high camp packed on its top from completely turning over. It was the most dangerous leg of our journey!

I returned home to enjoy ten days with mother who carefully watched me pack away my climbing gear just before she departed to an even higher place than I had just been.

76

TESTING HEARTS AND MINDS DUE TO BEING 'TOO GOOD TO BE TRUE'

The publication of this book will be the cause of more elation but also generate new opposition to the discovery of Noah's Ark and to the bold teachings of Scripture Science which are necessary for understanding the discovery. This is to be expected because anything that might change the world ought to be tested as the gospel itself has been tested from its earliest days. From the resurrection of Jesus, everything that has changed the world including every revival of the Christian faith have been controversial. However, opposition is greatest when something is "too good to be true."

Indeed, we learn to be cautious about things that seem too good to be true. 'Too good to be true' is a flag for pyramid schemes, for swindlers, and in some cases, relationships! Moreover, there have been many announcements of the discovery of Noah's Ark that did prove too good to be true. The more that hopes are built up by a discovery, the more disappointed will be believers in announcements that prove too good to be true. To keep from being hurt, disappointed, or appear foolish we learn to guard our hearts. Ordinarily, this is the wise thing to do. However, counterfeit claims exist because there is potentially something authentic to counterfeit. When something is 'too good to be true' but has great importance, we should carefully examine the evidence and the character, wisdom, and reputation of those who proclaim it.

Keep in mind that for a cynical world, the gospel is too good to be true. The gospel has long been long established, but now the recent discovery of Noah's Ark is again "too good to be true". So is understanding origins by using the light of the Bible thereby turning upside down almost everything now being taught in both mainstream and Creationist Science. That this discovery has come to light at just the time when it seems that the whole world including America are turning against biblical faith is just unbelievable!

The gospel tests hearts. This discovery also tests hearts! Due to my investigation, I had more reason for believing this discovery than others. Even so, however much I had investigated this discovery, however much I understood it, and however convinced that the discoverers were telling the truth, I knew that no amount of investigation can answer every question. Thus, I had to decide whether to associate my years of work and name with this controversial discovery. The truth is, I made my decision by the witness of the Holy Spirit, which born again believers know to be the spirit of truth. Those announcing the discovery did so in the transparent manner of people who are telling the truth.

What makes it inevitable that the whole world will hear of this discovery is that almost everyone from small children to the most erudite college professor have heard of Noah's Ark whether they reject it as a fable or believe the story as do small children. Because the Internet is becoming ubiquitous, even places in the world which have yet to hear of either Jesus or Noah's Ark may now hear of both. Such is the significance of this discovery that every responsible adult will eventually have to decide whether this discovery is authentic and if so, consider the implications.

Those few who maintain an open mind about such things are to be admired. However, after being confronted by someone we know or trust who believes and proclaims the discovery, we may have to decide for ourselves whether it be based on our own investigation or on trusting others. This discovery tests hearts because it is 'too good to be true.' One's decision ought to be accompanied by prayer and by trusting the Holy Spirit who dwells in circumcised hearts.

Those who have long been involved with this discovery soon forget the controversy as they come to understand its truth and associate with others who also understand its truth. However, I warn everyone who joins in supporting this discovery to expect opposition from some or many of their friends, associates, and even their own families. This end time discovery is not so different from the beginning of the gospel, even to the frivolous charge of it being a hoax. As the case of the Apostles, some of those who proclaim this discovery as Noah's Ark will probably have to pay a price for believing something which to the world appears too good to be true.

77

The Price to be Paid for the Freedom to Pursue and Proclaim Truth

The more that one learns about this discovery of Noah's Ark and Flood, the more excited that he or she will be, especially Christians who love the Bible. Most will wish to share their excitement with family, friends, and colleagues. But know that however great the reward of knowing and proclaiming truth, it may be costly to proclaim and teach it. Pioneering and proclaiming world-changing truth has always been costly.

One may ask why the advancement of truth should be costly and what cost those who pioneer world-changing truths should expect. It is costly because world-changing truth disrupts established authority incurring opposition from those who love and prefer their current status and privileges that may be impacted by this discovery. It will frighten some who know and respect you but do know not how to respond to such a message. Be patient and understanding because these at least comprehend the significance.

We are however inclined to believe what other people believe, especially those in our circles of acquaintance. As of this writing, few know and believe this discovery of Noah's Ark and Noah's Flood. Thus, the first and costliest price will be due to opposition from the same. This is because people expect world-changing truths to come from established authorities and not from ordinary people as the case of most of us.

Though the truth which sets us free is freely given to the recipients of God's grace, a price has to be paid for the freedom to pursue and proclaim truth. Some are fortunate to live in a country with a history of freedom of religion, freedom of speech, and freedom of assembly. For this, our spiritual ancestors paid the cost for following Jesus Christ who paid the first and greatest cost. Many suppose that our freedoms have been won and preserved due to military service and sacrifice. As noble as that may be and as necessary as it may have been at times, from the beginning of time men have sought to obtain or preserve their freedom through fighting and sacrifice of their lives. Yet, none of these truly won their freedom. Rather, the truth that makes us free begins with knowing and learning to obey Jesus Christ. This present truth was built on the same foundation.

Dr. Geisler recognized that the Ark projected needed institutional support. But one of the two seminaries that had co-founded was allied with Old Earth Creationists while the other strongly embraced the view of the Young Earth Creationists. The entire evangelical world has been built around understandings that are incompatible with either a worldwide Flood or an old earth. We traveled to meet the leader of one Creationist organization with whom we hoped that we might somehow be able to work. However much the mutual respect, our different understandings of the worldwide Flood made that difficult, at least for the time.

As we were driving home, I shared with Norm that I lacked both an organization and the funds by which I might create such an organization. I also lacked Dr. Geisler's valuable name and reputation in the Christian world. It then occurred to me that I did have one advantage, my freedom! Dr. Geisler responded, "Yes, and I envy you!'

I understood why Dr. Geisler responded this way. Though many advantages and good things come from our relationships, obligations, and associations, even our families, any or all of these may wish to limit our freedom to pursue truth. Thus, the Apostle Paul explained how being single allows one's time and resources to be completely devoted to Christ.[73] Being single, I had no family to suffer from my devotion to this work. But our memberships and associations also constrain our freedom should our pursuit of truth conflict with some of these. These might be the churches or denominations to which we belong, but especially our positions in various organizations whether secular or religious. I have had to limit my involvement with various organizations and in some cases, companies and organizations that I was instrumental in establishing.

---

73 1Corinthians 7:26-27

Jesus expects his flock to forsake all to follow him. Thus, pursuing and proclaiming truth will be costly. If not, one has been pursuing something other than truth. However, those who do forsake all to follow truth will experience the fellowship of the Holy Spirit, God's most wonderful gift in the present age.

It is also true that our memberships and associations bring accountability. With regard to truth and righteousness, accountability is essential. But such accountability might not always be accountability to truth or righteousness, at least with regard to all things. Patronage or the freedom to participate are among the advantages of our associations. The price we must pay to be free may be that of forfeiting such advantages whether they come from the secular government or associations or from religious connections.

When I decided to support the recent discovery of Noah's Ark, I knew that I was putting my scholarship and reputation at risk by associating it with this discovery. So did Norm Geisler. Indeed, we should be held accountable for whatever we claim or endorse. It has cost me platforms and opportunities to share my understanding of the discoveries of Noah's Ark and Flood. That immediately included an opportunity for promoting my work by a producer who had promoted a rival Ark site in a famous CBS documentary.

When Norm Geisler endorse this discovery, he also upset many of his colleagues. Dr Geisler however was already famous for forfeiting patronage and associations on account of truth. Notwithstanding how much we may have to stand alone on account of either righteousness or truth, if we belong to Jesus Christ, we are never truly alone as we are comforted by his Holy Spirit. Even so, we should never seek to escape accountability to Jesus Christ and to the body of Christ at large. If we claim to be teaching the truths of Christianity, we are accountable to every member of the body of Christ!

78

NOAH'S ARK AND THE KINGDOM OF GOD

The gospel of Jesus Christ has not just one but two messages of good news. The first is to individuals, the forgiveness of our sins through the atoning blood of Jesus Christ which leads to Eternal Life. This promise of Heaven awaits the faithful. The second is a corporate message, the promised Kingdom of God. The New Testament begins with John the Baptist then Jesus telling his disciples to proclaim that the Kingdom of God is near. Alternately referred to as the Kingdom of Heaven, this is the promise of God's direct rule of the people of the earth as promised in the Old Testament Prophets. This promised Kingdom began with the Church on the Day of Pentecost which however is separate from the governments and nations of the world. In the present age, members of the visible church serve as ambassadors of Christ's invisible Heavenly Kingdom which will one day rule all the earth. Noah serves as a type of such ambassadors while he was preaching and building the Ark.

We are rightly inclined to associate the first gospel message with Jesus's First Coming and the second message with his Promised Second Coming. Because Jesus compared his Second Coming to the Days of Noah, the historical Noah's Ark, which has just been discovered, serves as a type of the Second Coming with a new beginning of the world that accompanies his Second Coming. However these messages are not entirely separated by the two Comings. Jesus gave us parables pointing to the Kingdom of God as growing in the earth between his First and Second Comings. In Chapter 68, I reviewed the many ways this has been happening. This comes about through the joyful suffering and service of the servants of Jesus and results in their leadership in the Kingdom of God accompanying the Second Coming. The Apostle Paul encourages the Church at Thessalonica with this message concerning their persecution for the name of Jesus:

All this is evidence that God's judgment is right, and as a result you will be counted worthy of the kingdom of God, for which you are suffering. God is just: He will pay back trouble to those who trouble you and give relief to you who are troubled, and to us as well. This will happen when the Lord Jesus is revealed from heaven in blazing fire with his powerful angels. He will punish those who do not know God and do not obey the gospel of our Lord Jesus. They will be punished with everlasting destruction and shut out from the presence of the Lord and from the glory of his might on the day he comes to be glorified in his holy people and to be marveled at among all those who have believed. This includes you, because you believed our testimony to you. [2Thess 1: 5- 10]

The Kingdom of God in America

Though much neglected by evangelicals today, Christians have not always ignored the message of the Kingdom of God. This was particularly the case of Christians working across denominations in the early revivals of America. The Coming Kingdom was rightly associated with righteous social progress including democracy, the elimination of slavery and child labor, women's suffrage, and proclamation of the gospel to the nations. This vision of the Kingdom of God was also associated with then Christian America. This optimistic vision of the future informed the secular vision of America whose optimism has long contrasted with European cynicism.[74] Regardless America's present money-driven economy and corrupt secular government which function as a parallel world-dominating Babylon, America in much of the world continues to be identified with Christianity.

In Chapter 68, I explained how this vision of the Coming Kingdom without separating God's people from the kingdoms of this present world including America's money-driven society changed into liberal Christianity where progress came to be associated with scientific progress and a socialistic and humanistic version of the Kingdom of God, even incorporating the scientific challenge to the truth of the Scriptures and the liberal criticism of the Bible. Thus arose in reaction the Fundamentalist movement which after WW II would be called *evangelicalism*. Evangelicals turned from looking forward to the Kingdom of God with the Coming of Christ to looking for an antichrist to rule the world. Expecting to be raptured to Heaven before the appearance of the antichrist, evangelicals forgot about the Kingdom of God and also America. This is especially sad for children of believers

---

[74] c.f. H Richard Niebuhr, *The Kingdom of God in America* (1937, Harper & Row)

because there is no marriage in Heaven. The earth is for the children as also for believers spiritual growth. Do not even those in Heaven look forward to returning with Jesus to the earth![75]

Having given up on seeing the Kingdom of God appear, many evangelicals minimize or dismiss addressing the scientific challenge to the Bible in favor of simply the preaching of the gospel. Unfortunately, the unanswered scientific challenge to the Bible has limited openness to the gospel. This is why the faith has been growing mostly outside America and the West. Even the converted soon find themselves needing to address the scientific challenges to the Bible.

We may also forget that apart from the biblical understanding of origins, the gospel may not even make sense. Thus, the gospel of John begins by explaining Jesus as the Creator of all things. His redemptive blood is due to the problem of sin which began with Adam. The Jesus upon whom our salvation depends mentioned things recorded in the Genesis account of origins including Noah's Flood more than anyone else in the Bible. Thus, apart from good answers to the scientific challenge to the Bible, new believers who have been trained in the narrative of mainstream science will be confused about origins as was once the case with me. That makes us less effective not only in sharing the gospel, but also as believers. Moreover, the schools will still be teaching unbelief to our children. America will never return to being a Christian nation unless we effectively address the matter of origins. About this, my friend Ken Ham heartily agrees.

As I will explain below, we can now look forward to teaching origins as a matter of general education serving as preparation for the gospel of salvation.

Naturalistic Science as the Powerful Delusion Sent by God

The key New Testament Scripture for supposing that a future antichrist must precede the Coming of Jesus is in the second chapter of Paul' second letter to the Thessalonian church.

Concerning the coming of our Lord Jesus Christ ... Don't let anyone deceive you in any way, for that day will not come until the rebellion occurs and the man of lawlessness is revealed, the man doomed to destruction. He will oppose and will exalt himself over everything that is called God or is worshiped, so that he sets himself up in God's temple, proclaiming himself to be God. Don't you remember

[75] c.f. Revelation 19:14

that when I was with you I used to tell you these things? And now you know what is holding him back, so that he may be revealed at the proper time. For the secret power of lawlessness is already at work; but the one who now holds it back will continue to do so till he is taken out of the way. And then the lawless one will be revealed, whom the Lord Jesus will overthrow with the breath of his mouth and destroy by the splendor of his coming. The coming of the lawless one will be in accordance with how Satan works. He will use all sorts of displays of power through signs and wonders that serve the lie, and all the ways that wickedness deceives those who are perishing. They perish because they refused to love the truth and so be saved. For this reason ***God sends them a powerful delusion*** *so that they will believe the lie and so that all will be condemned who have not believed the truth but have delighted in wickedness. [2Thessalonians2: 1-12]*

There is no mention of the word *antichrist* in this passage, but whoever is the man of lawlessness would qualify as being that which most understand as the antichrist. We should however consider a corporate Man of Lawlessness (often translated as 'Man of Sin') as an antitype of Christ's Church, the corporate body of Christ. The corporate Man of Lawlessness are "Christians" proclaiming man himself as being over everything called God, exactly the case of today's secular and scientific world which, as I have explained, originated from the state-controlled churches of Europe. This was something new in the history of mankind, which had previously held to some type of religion.

The question that the Apostle Paul is answering in this passage concerns what is holding back the return of Christ, not what is holding back the antichrist as most have taught this passage. Moreover, those who see this as holding back the antiChrist ignore the fact that in the New Testament, God's Temple is the church, the body of Christians from which came those who would embrace the scientific challenge to the Bible! Early Christians attracted to Gnostic teachings were precursors of those who today trust in Science. By trusting in Science, man places himself above all things called God including the inspired Scriptures. Who can deny that there has ever been a display of power and wonders advertised as having challenged and defeated God as the case of modern science! The higher critics of the Bible used this very argument!

Fortunately, this passage also assures us that the pretentions of naturalistic science will be overthrown by the 'breath of his mouth,' that is the inspired word of God at the splendor of Jesus's Coming.

According to this reading, and as I mentioned in Chapter 71, God himself sent this powerful delusion of naturalistic Science. However, the scientific challenge to the truth of the Bible freed the body of Christ from the power of the state as exercised through institutional Christianity. If the discovery of Noah's Ark is a sign of his Coming, we can also recognize the Scriptures as breathed out by God and the splendor of his Coming as soon defeating the scientific challenge to the Bible. The same may also serve as the means by which Jesus Christ gathers believers who have a direct relationship with him so as to be a truly great light to the nations.

79

How Believers Can Show the World How to do Science with Excellence

One thing helping launch the Scientific Revolution were questions relating to various scientific matters for which honors and prizes were offered to whomever could provide the best answers. While I have neither silver nor gold to offer as prizes, I promise to honor those who can provide answers or good responses concerning matters covered in this book as I list below:

Eight Questions for Young and Old Earth Creationists

(1) Why do erratic stones as those I found at the Big Rock Nature Preserve in Charlotte, NC (see Fig 25, Chapter 25) supposed deposited by Ice Age glaciers exist so far south of the southern extent of the Ice Age? Are these Ice Age rocks commonly found in lands supposed absent of Ice Age glaciers? How many of the latter can we identify, locate, and photograph? May we map these stones to get a better understanding of the changes in the earth due to the sudden and powerful release of the antediluvian underground springs at the time of Noah's Flood.

(2) What is a sound geological explanation for the placement of these erratic stones so far from their place of origins? Can the placement of these rocks be explained other than having been propelled by the release of the antediluvian underground springs?

(3) What physical processes could have propelled supposed Ice Age glaciers along level ground, even over mountains? How do glaciers push rocks as large as small houses for tens of miles across frozen earth? If glaciers do somehow move along level ground what keeps

them from moving over and around instead of pushing these heavy rocks?

(4) Whether Pleistocene megafauna such as mammoths, saber-toothed tigers, and cave bears supposed as becoming extinct at the end of the Ice Age be better understood as the remains of animals made extinct by Noah's Flood?

(5) Whether the prehistoric graves from the Middle Paleolithic to the Early Bronze Age that are currently explained as elaborate religious burials be better understood as the dwellings of antediluvian civilizations ended by Noah's Flood?

(6) What prevents the earth's ionosphere which moves with the turning earth from producing the earth's north/south magnetic field? Should there be a correlation between the different planets' rates of rotation, the magnitude of their ionospheres, and the strength of their magnetic fields?

(7) Is there is a significant correlation between Jupiter's orbit around the sun and the shifting of the direction of the sun's magnetic field? If so, how might that relate to sunspot activity and periodic changes in the earth's climate?

(8) Is the recent discovery on Mount Ararat Noah's Ark? Here, one should respond to claims made in this book because of all those involved in this discovery I understand the discovery best and make the strongest claims.

Most mainstream scientists will not have the motivation to engage in such a project until a new biblical paradigm attracts large numbers of disciples and students. However believers should look forward to discovering whether there exists a biblical school of origins that is manifestly superior to mainstream science. An excellent way to do science is for Young and Old Earth Creationists to join with me in providing answers to these 8 scientific questions which will also interest mainstream scientists.

Those willing to do research or write well-reasoned papers concerning any of these should submit them to any journal that might be interested, whether of mainstream or Creation Science. I may consider such papers for the new online *Journal of Scripture Science,* or else review them in this new journal.

Even better will be apologetics conferences addressing these questions, which of course will have the interest of the world. Here is an opportunity for Creationists with different points of view to work together in a friendly way to show the world how Christians forthrightly and honestly confront and resolve conflicting scientific views.

Many years ago, Dr. Geisler tried to get the leader of the most prominent Young Earth Creationist organization to work with the leader of what is now the largest Old Earth Creationist organization to develop a common evangelical understanding of origins. He did not then succeed. These Creationist leaders with different views of origins built separate but competing organizations. Dr. Geisler however never lent his name and scholarship to either school as he would do in the case of Scripture Science. Nor did he ever endorse any claim for the discovery of Noah's Ark until this recent discovery.

As was the case of Dr. Geisler, Scripture Science has something in common with both schools of origins, these being the worldwide Flood of one school and the old earth view of the other school. I value the respective scientific contributions of both schools of origins. I trust that those belonging to these schools value what has been presented in this book. The leaders of both organizations do understand the potential of the discovery on Mount Ararat to change Christian understandings of origins. I regard them as friends and Christian brothers if we hold different understandings of origins and see these differences as perhaps vital to the future of our faith. Whatever the result of these discussions, the leaders and advocates of these different schools of origins together with the scientists in the various fields of mainstream science will obtain a deeper understanding of the history of science and the history of origins. Regardless their conclusions, I will continue to regard as Christian brothers and sisters all who hold to the exclusive salvation by Jesus Christ and continue to regard the Bible as the exclusive foundation of our common faith.

80

HOW CHRISTIANS TEACHING A BIBLICAL UNDERSTANDING OF THE PAST WILL CHANGE THE WORLD

The book has explained the truth concerning the past from the origins of mankind and the world to the discovery of the world's historical nations by European explorers. However the significance of understanding the truth about the past concerns the future.

"Who controls the past controls future. Who controls the present controls the past" [George Orwell, 1984, Book 1, Chapter 3, p34)

Regardless of who controls our current understanding of the past, it is the God of the Bible who controls the future. No matter who controls the academy, the schools, the media, or today's governments, those who read, study, and understand this book will learn the true story of the past thereby giving Christians a bright and hopeful future regardless of what troubles beset our current age. Bringing about a great and wonderful change in the world requires only that Christians work together to learn and teach this new understanding of the past from the light of the Scriptures to our fellow believers, to our children, and to any who might be interested.

Note that this can be done independently of political involvement! This is not to deny that believers should support whatever political changes may be wise and good with our voices and votes including opening schools to the truth concerning the past. But instead of spending so many resources fighting the darkness, all that believers need do is turn on the light concerning the past! This light will get attention of the world and thereby change the world.

In this regard, we should commend both Young and Old Earth Creationists for building scientific communities separate from that of mainstream science. While neither the Old nor the Young Earth versions of Creation Science have threatened mainstream science, Scripture Science has two great advantages. The first is its exciting association with the recent discovery of Noah's Ark which will bring Scripture Science's understanding of the past to the attention of the world. The second is the embarrassing revelation that the Ice Age has simply been a misunderstanding of Noah's Flood. As this understanding grows among Christians, mainstream scientists will be forced to explain why they misunderstood the evidence of something so great as the worldwide Flood of Noah. The more that scientists and scholars address the matter, the more embarrassing this will become. Holding to the current view will seem ever more ridiculous. No view that appears ridiculous can long survive.

Practitioners of mainstream science who seek truth will soon convert to the biblical view of origins. It is however unnecessary that all who trust mainstream science do so, as has been the case of previous paradigm shifts. Regardless the evidence, some do not wish to believe the Bible and the gospel even if these be true. As we are free to believe the gospel and the Scriptures, others are free not to do so. We who seek the truth concerning origins do so because of our commitment to truth and to advance the Kingdom of God.

Keep in mind that the scientific challenge to the Bible came about when professional scientists worked together to develop a naturalistic understanding of the world while marginalizing those committed to the biblical understanding of origins (Chapter 22). Christians can now marginalize those committed to the mainstream scientific narrative but fail to acknowledge the supernatural but historical Flood of Noah.

Though we invite professional scientist to investigate this matter, and however much this discovery along with Scripture Science may benefit from their expertise, they are not to be trusted as the ultimate judges. For how can those who failed to recognize the manifest evidence of Noah's Flood be trusted for such an important undertaking? On the other hand, believers do have an obligation to do what we can to promote and teach the truth concerning the past which can and will change the world.

ANTEDILUVIAN AND POST-DILUVIAN ANCIENT AMERICA

It is not just the world's understanding of the ancient past that will be affected by this change because the mistaken understanding of the earth's past resulted in the shift in understanding the world and America's past from events in Christian history to the modern and secular understanding, beginning with the Enlightenment until the most recent revisionist history of America that is widely taught today and believed even by Christians.[76]

As I explained in Chapter 35, racism and the adaption of the classical world's racial justification of slavery was a product of the European creators of the Bible-rejecting Enlightenment. As American slave owners and philosophers adopted the racism of the eighteenth century Enlightenment, in the later nineteenth century American industrialists and scientists introduced the anti-Christian racist teachings of Social Darwinism. Though Marxist cannot be faulted for pointing out injustices in American history, Marxist share the Christian and Bible-rejecting view of those responsible for these evils yet have nothing to offer as a solution but a system responsible for even greater evils including their attempts to erase the truth of history.

This Marxist-driven revisionist history uses the evidence of a highly populated antediluvian America as evidence that the United States is a product of the genocide of native Americans. Revisionist historians teach a literal and cultural genocide of America's natives by Christian European settlers. These estimates however are much based on America's advanced and considerable antediluvian population. The earliest European explorers discovered that native Americans knew about the Flood but also denied knowing who were the violent people in the massive burial mounds of pre-Columbian America.

Studying the map of migrations in Fig 9 of Chapter 7, we can see that native Americans are descendants of Noah's son Shem. Navigating by the sun from the coast of Asia, they landed in Ecuador from which place the Formative civilizations of America derive. Around the time of Jesus, descendants of these first Americans traveled up the Mississippi and Ohio Rivers to become the Mississippian culture of the earliest American natives.

Due to the great forest then covering Eastern America, native settlements were limited to the rivers and coasts. The same was true of the first European settlers. But the European setters had iron axes by which they could clear the forest thereby increasing the available land and productivity more than a hundred fold both for people and animals. By depending on agriculture rather than hunting and finishing, America was able to support far more people and animals. Urbanization

---

[76] c.f. Howard Zinn, *A People's History of the United States* (1980, Harper Collina)

notwithstanding, there is even more wildlife in Eastern America today than when the first Europeans arrived. This is due to America's productive gardens and fields. Contrary to the beliefs of Nature-worshipping environmentalists, God-fearing humans make the earth ever more beautiful and productive.

This is not to deny that European settlement of the New World began as colonial projects that exploited natives as has been the case of most of the expanding settlements in both the Old and New Worlds from ancient times until today. This was also the case among the native populations of America. Aside from the influence of Christianity and the God of the Bible, the strong have exploited the weak throughout human history.

What these revisionist histories overlook is how much the Unites States of America has been a notable exception to this rule. This was due to the Christian purposes of most of the earliest American settlements. What made the 13 colonies exceptional even from other English settlements in the New and Old Worlds is that mostly forested Eastern America offered religious freedom to those willing to face the dangers and hardship of living in this forest which was so remote from the comforts of the Old World. Some of the earliest and most influential colonists hoped to convert rather than exploit the natives. But regardless their hopes of sharing the gospel, these Christians aimed and did generally treat the natives with respect and justice. Of course, not everyone did so, especially those with money and power as has been and will continue to be the case until America becomes a truly Christian nation. For such an end, America has truly been exceptional leading to the blessings that have come to this nation.

LET US WORK TOGETHER TO HELP ALL MANKIND DISCOVER OUR ORIGINS

Keep in mind that neither Old nor Young Earth Creationists have enough light on the historical past to provide a true history of our beginnings and certainly not such as to change the world. Old Earth Creationists suppose that the antediluvian civilizations lie buried some place in the Ancient Near East. They have not even inspired anyone to go looking for these ruins. But Young Earth Creationists tell us that the Flood vaporized even the evidence of the drowned antediluvian civilization. Thus, neither Old nor Young Earth Creationists can trace our ancestors from the Ark that I visited near the top of Mount Ararat. Nor can their understanding of the past deliver such a bright and hopeful future. Come now brothers, let us work together to reveal the history of all mankind, thereby become a light to which all nations will come.

Topical Bibliography

Some will wish to know some of the books, papers, and other sources used for understanding the history, teachings, and practices of the historical sciences. This is by no means an exhaustive list of the far greater number of sources that I have used in more than thirty years of researching these topics, but these texts are useful for learning and understanding more about the various scientific fields addressed. I include notes concerning some of the more useful entries.

The History and Theory of Archaeology

William F. Albright, *From the Stone Age to Christianity* (1957), an erudite summary of archaeological prehistory up to the middle of the twentieth century

V. Gordon Childe, *The Aryans* (1926), *Man Makes Himself* (1951), *New Light on Most Ancient East* (1953), *The Dawn of European Civilization* (1957)

Glyn Daniel, *The Idea of Prehistory (1963), One Hundred Years of Archaeology (1950),* two standard texts for the history of archaeological prehistory

Donald K Grayson, *The Establishment of Human Antiquity (*1983), a rare history of this subject.

Marvin Harris, *The Rise of Anthropological Theory* (2001)

J.P. Mallory, *In Search of the Indo-Europeans: Language, Archaeology, and Myth* (1989)

Colin Renfrew, *Before Civilization* (1973)

Bruce G. Trigger, *A History of Archaeological Thought (1989)*

David Wilson, *The New Archaeology* (1974)

Archaeological and Anthropological Texts

Ruth Amiran, *Ancient Pottery of the Holy Land: From its Beginnings in the Neolithic Period to the Iron Age* (1963,1969)

Amnon Ben-Tor, Editor, *The Archaeology of Ancient Israel* (1992)

Charles Burney, David Lang, *The People of the Hills: Ancient Ararat and Caucasus (1971)*

Michael D. Coe, *The Maya* (1993), *Mexico: from the Olmecs to the Maya* (1993), discusses connections between Asia and the Americas

Arthur Cotterell, Editor, *The Penguin Encyclopedia of Ancient Civilizations* (1980)

Harriet Crawford, *Sumer and the Sumerians* (1991)

Charles Darwin, *The Descent of Man, and Selection in Respect to Sex* (1871)

I.E.S. Edwards et al, Editors *The Cambridge Ancient History* (Vol 1, Part I, 1970)

Robert W. Ehrich, Editor, *Chronologies in Old World Archaeology (Vol I and II, 1992)*

Stuart J. Fiedel, *Prehistory of the Americas,* Second Edition (1992)

James A. Ford, *A Comparison of the Formative Cultures of America* (1969)

Jacquetta Hawkes, *The Atlas of Early Man (1976)*

Kathleen Kenyon, *Digging Up Jericho* (1957)

Samuel Noah Kramer, *The Sumerians* (1963)

Sir John Lubbock, *Pre-Historic Times* (Sixth Edition, 1904), *The Origin of Civilization and the Primitive Condition of Man* (1870)

Charles Lyell, *The Geological Evidence of the Antiquity of Man* (1863, 2004 Dover Edition)

Patty Gerstenblith, *The Levant at the Beginning of the Middle Bronze Age (*1983)

Betty J. Meggers, Clifford Evans, and Emilio Estrado, *Early Formative Period in America: the Valdiva and Machalio Phases* (1965)

Samuel George Morton, with additions by Louis Aggasiz et al, *Types of Mankind Or, Ethnological Researches Based Upon the Ancient Monuments, Paintings, Sculptures, and Crania of Races, and Upon Their Natural, Geographical, Philological, and Biblical History* (1855)

Hans J. Nissen, *The Early History of the Ancient Near East: 9000-2000 BC* (1988)

Konstantine Pitskhelauri, URUK MIGRANTS IN THE CAUCASUS, *Bulletin of the Georgian Academy of Sciences,* vol 6, no. 2, 2012, pp 153-161

Joseph Prestwich, *Address Delivered at the Anniversary Meeting of the Geological Society of London* (1872)

Michael Roaf, *Cultural Atlas of Mesopotamia and the Ancient Near East* (1990)

Mitchell S. Rothman, *Uruk, Mesopotamia, and its Neighbors* (2001)

Adam T. Smith, Karen S. Rubinson, *Archaeology in the Borderlands: Investigations in Caucasia and Beyond* (2003)

Dean Snow, *The Archaeology of North America* (1976), includes Glacial Kame, Hopewell comparison

Margreet L. Steiner, Ann E. Killibrew, Editors, *The Archaeology of the Levant: c: 8000-332BCE* (2014)

W. J. Sollas, *Ancient Hunters and their Modern Representatives* (1924)

Sir Leonard Woolley, *Excavations at Ur* (1954), Wooley describes his excavation of the Royal Tombs or UR

THE ICE AGE

Alastair G. Dawson, *Ice Age Earth: Late Quarternary Geology and Climate* (1992)

Sir John William Dawson*, The Meeting Place of Geology and History (1893), The Canadian Ice Age: Being Notes on the Pleistocene Geology of Canada, with Especial Reference to the Life of thee Period and It's Climatic Conditions* (1893)

Luna B. Leopold, M. Gordon Wolman, John P. Miller, *Fluvial Processes in Geomorphology* (1964, 1992)

Andrew Goudie, *Environmental Change* (1977)

Sir Henry H. Howorth, *The Glacial Nightmare and the Flood* (Vol I, 1892; Vol II, 1893), *Ice or Water* (Vol I, 1905)

John Imbrie and Katherine Palmer Imbrie, *Ice Ages (1979),* history of Ice Age theory

Doug Macdougall, *Frozen Earth: The Once and Future Story of Ice Ages* (1994)

W.S.B. Paterson, *The Physics of Glaciers, Second Edition* (1981) The History and Distribution of Fauna

Edwin H. Colbert, *Wandring Lands and Animals: The Story of Continental Drift and Animal Populations* (1985, Dover)

Charles Darwin, *The Origin of Species by Means of Natural Selection or The Preservation of Favored Races in the Struggle for Life* (1859)

Chet Van Duzer, *Floating Islands: A Global Bibliography* (2004)

Sir Henry H. Howorth, *The Mammoth and the Flood (*1887)

Ernest Mayr, *Systematics and the Origin of Species from the viewpoint of a zoologist* (1964, Dover)

Peter D. Ward, *Rivers in Time: The Search for Clues tot he Earth's Mass Extinctions* (2000, Columbia University Press)

The History and Theory of Geology

L.H. Ahrens, *Distribution of the Elements in Our Planet* (1965)

Robert Bakewell, Benjamin Silliman, *An Introduction to Geology* (1839)

Sir John William Dawson, *The Meeting Place of Geology and History* (1894)

Archibald Geikie, *The Founders of Geology* (1891)

Charles Coulston Gillipsie, *Genesis and Geology (1951),* the standard text for how the uniformitarian view triumphed over diluvium and related views

A. Hallam, *Great Geological Controversies,* Second Edition (1989)

Stephen L. Harris, Agents *of Chaos:Earthquakes, Volcanoes, and other Natural Disasters* (1990, Mountain Press)

Luna B. Leopold, W. Gordon Wolman, John P. Miller, Fluvial Process in Geology (1992, Dover)

Charles Lyell, *Principles of Geology* (Volumes I,II, and III, 1830-1833)

J.D. MacDougall, *A Short History of Planet Earth: Mountains, Mammals, Fire, and* Ice (1996, Wiley)

Alexander R. McBirney, *Igneous Petrology: Second Edition* (1993)

Joseph Prestwich, Address *Delivered at the Anniversary Meeting of the Geological Society of London* (1872)

H.H. Read, *The Granite Controversy* (1957)

Martin J.S. Rudwick, *The Meaning of Fossils* (1976), *Bursting the Limits of Time* (2005), *Men Before Adam* (2008), *Earth's Deep History* (2014), the most comprehensive geological history for the topics covered in the present book

Alfred Wegener, *The Origin of Continents and Oceans* (1929)

The History of Ideas

Tuska Benes, *In Babel's Shadow: Language, Philology, and the Nation in Nineteenth-Century Germany (2008)*

James E. Force, *William Whiston: Honest Newtonian* (198**5)**

John Fauvel,et al, editors, *Let Newton Be!: A new perspective on his life and works* (1988, Oxford)

Tom Holland, *Dominion: How the Christian Revolution Remade the World* (2019, Basic Books).

Arthur O. Lovejoy, *The Great Chain of Being* (1964)

Suzanne L. Marchand, *German Orientalism in the Age of Empire: Religion, Race, and Scholarship* (2009), *Down from Olympus: Archaeology and Philhellenism in Germany, 1750-1970* (1996), WHERE DOES HISTORY BEGIN: J.G. HERDER AND THE PROBLEM OF NEAR EASTERN CHRONOLOGY IN THE AGE OF THE ENLIGHTENMENT, *Eighteen-Century Studies,* vol. 47, no. 2, (2014) pp. 157-175

Alister E. McGrath, *Reformation Thought (1993)*

William Paley, *Natural Theology (1802)*

Richard H. Popkin, *The History of Skepticism from Savonoarola to Bayle* (2003); *Isaac la Peyrere (1596-1676) His Life, Work, and Influence* (1987, Brill), *Disputing Christianity* (2007, Humanity Books)

Dorothy Ross, *The Origins of American Social Science* (1991)

Paolo Rossi, T*he Dark Abyss of Time*, translated by Lydia G. Cockrane (1984)

Jonathan Sheehan, *The Enlightenment Bible* (2005)

Stephen D. Snobelen, ISAAC NEWTON, HERETIC, THE STRATEGIES OF A NICODEMITE, *The British Journal for the History of Science*, (December 1999), pp 381-419

Peter Toon, *Puritans, the Millennium, and the Future of Israel* (1970)

James Turner, *Without God, Without Creed: The Origins of Unbelief in America* (1985)

NOAH'S ARK AND FLOOD

Johannes Buteo, *The Shape and Capacity of Noah's Ark* (1554), translated by Tim Griffith and Natali Miller

John Warwick Montgomery, *The Quest for Noah's Ark* (1972), covers historical reports of Noah's Ark on Mount Ararat

Ronald L. Numbers, *The Creationists* (1992), the origins of the current versions of young earth Creationism

George McGready Price, *Evolutionary Geology and the New Catastrophism* (1926,Pacific Press)

Michael J Oard and John K Reed, *How Noah's Flood Shaped our World* (2017)

Alfred H Rehwinkel, *The Flood in the Light of the Bible, Geology, and Archaeology* (1951)

John C Whitcomb, Henry M. Morris, *The Genesis Flood: The Biblical Record and its Scientific Implications* (1961, Presbyterian and Reformed)

Ellen G. White, *Patriarchs and Prophets* (2002, Interamrican Publishing Association)

Philip Ernest Williams, *The Archaeological Evidence of Noah's Flood* (2011), *Ancient Light from this Mountain* (2022)

Davis A. Young, *The Biblical Flood: A Case Study of the Church's Response to Extrabiblical Evidence* (1995)

QUATERNARY THEORY AND SCIENCE-BASED DATING

M. J. Aitken, *Science-based Dating in Archaeology* (1990)

Richard B Alley, *The Two-Mile Time Machine: Ice Cores, Abrupt Climate Change and Our Future* (2000), covers attempts to find a chronology in ice cores

M.G.L Baillie, *Tree Ring Dating and Archaeology* (1980), *A Slice Through Time: dendrochronology and precision dating* (1995)

Raymond S. Bradley, *Paleoclimatology: Reconstructing Climates of the Quaternary* (1999)

J.K. Charlesworth, *The Quaternary Era with Special Reference to its Glaciation (Vols I and II,* 1957)

Thomas Cronin, *Principles of Paleoclimatology* (1999)

Andrew Goudie, Environmental Change (1977, Clarendon Press)

Lucille Lewis Johnson, *Paleoshorelines and Prehistory* (1992, CRC Press)

Richard Foster Flint, *Glacial and Quaternary Geology* (1971)

Jay Stratton Noller, Janet M. Sowers, William R. Lettis, Editors, *Quarternary Geochronology: Methods and Applications* (2000, American Geophysical Union)

Claudio Vita-Finzi, *Recent Earth History* (1973)

Gunther A. Wagner, *Age Determination of Young Rocks and Artifacts: Physical and Chemical Clocks in Quaternary Geology and Archaeology* (1995, Springer)

Miike Walker, Quarternary Dating Methods (2005, Wiley)

Stephen Weiner, *Microarchaeology: Beyond the Visible Archaeological Record (2010)*

PLATE TECTONICS AND EARTH'S MAGENETIC FIELD

Claude Allegre, *The Behavior of the Earth: Continental and Seafloor Mobility* (1988, Harvard)

Stephen Blundelll, *Magnetism: A Very Short Introduction* (2012, Oxford)

Stephen L. Harris, *Agents of Chaos: Earthquakes, Volcanoes, and Other Natural Disasters* (1990, Mountain Press)

David M. Lawrence, *Ocean Floor Mapping and the Earth Science Revolution* (2002, Rutgers)

Ronald T. Merrill, Michael W. McElhinny, Phillip L, McFadden, *The Magnetic Field of the Earth: Paleomagnetism, the Core, and the Deep Mantel* (2008, Academic Press)

Ronald T. Merrill, *Our Magnetic Earth: The Science of Geomagnetism* (2010, University of Chicago Press)

David M. Lawrence, *Upheaval from the Abyss (*2002)

M. W. McElhinny, *Paleomagnetism and plate tectonics* (1973)

Eldridge M. Moores, *Shaping the Earth: Tectonics of Continents and Oceans* (1990, Scientific American)

Naomi Oreskes, *The Rejection of Continental Drift: Theory and Method in American Earth Science* (1999), *Plate Tectonics: an Insider's History of the Modern Theory of the Earth* (2003)

Gollian Turner, *North Pole, South Pole: The Epic Quest to Solve the Great Mystery of the Earth's Megnetism* (2010,2011)

ASTROPHYSICS, COSMOLOGY, AND COSMOGONY

John D. Barrow, *The Origin of the Universe* (1994, Basic Books)

James Binney, *Astrophysics, A Very Short Introduction* (2001, Oxford)

Giovanni Caprara, *The Solar System* (2001, Firefly)

Peter Coles, *Cosmology: A Very Short Introduction* (2001, Oxford)

G. Brent Dalrymple, *The Age of the Earth* (1991, Standford)

John Gribbin, *Galaxies: A Very Short Introduction* (2008, Oxford)

Robert Kandel, *Water from Heaven: The Story of Water from the Big Bang tot he Rise of Civilization and Beyond* (2003, Columbia University Press)

Andrew King, *Stars: A Very Short Introduction* (2012, Oxford)

Neil McBride and Hunter Gilmore, Editors, *An Introduction to the Solar System* (2004, Open University)

Steven Weinberg, *The First Three Minutes: A Modern View of the Origin of the Universe* (1977, Bantam)

About the Author

Philip Ernest Williams began his professional career as a systems engineer designing and overseeing the world's first digital and fiber optics networks for local exchanges. His last project was designing the network management system for the US Department of Defense. Williams combined his work as a systems engineer with the study of philosophy, working with some of the twentieth century's most notable philosophers and theologians including Charles Hartshorne, Richard Popkin, Harvey Cox, and Norman Geisler. Doing pro-bono work designing infrastructure projects in Haiti and Nigeria, Williams worked with Presidents, Prime Ministers, Governors, and development officials as well as the ocean scientist John Pina Craven. For the promotion of Christian unity and charity, he has worked with leaders across the Christian spectrum, briefly serving as the Director of the New Beginnings Inner City Ministry.

Following his retirement in 1990 from the telecommunications software company he founded, Williams applied his systems experience to the complex problem of human origins and populations, specifically whether there exists evidence of Noah's Flood. In 2011, he published the first ever account of the archaeological evidence of Noah's Flood. In 2014, he and the archaeological official Sheila Bishop became the first Americans to visit the ancient wooden structure near the top of Mount Ararat as discussed in this book. In 2022, he published *Ancient Light from this Mountain,* a guide for archaeological understanding of the remains of Noah's Ark on Mount Ararat. He currently serves as the Scientific Advisor to the Mount Ararat Discovery Foundation.

Books by the Author

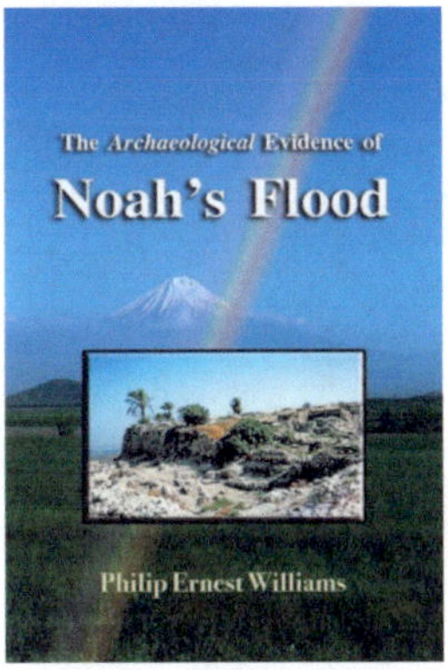

Available at **www.christianleadersandscholars.com**

Printed in France by Amazon
Brétigny-sur-Orge, FR

51790229R10192